THE WATER REUSE ROADMAP

WEF Special Publication

2018

Water Environment Federation
601 Wythe Street
Alexandria, VA 22314-1994 USA
http://www.wef.org

About WEF

The Water Environment Federation (WEF) is a not-for-profit technical and educational organization of 33,000 individual members and 75 affiliated Member Associations representing water quality professionals around the world. Since 1928, WEF and its members have protected public health and the environment. As a global water sector leader, our mission is to connect water professionals; enrich the expertise of water professionals; increase the awareness of the impact and value of water; and provide a platform for water sector innovation. To learn more, visit www.wef.org.

Prepared by the **Water Reuse Roadmap**
Task Force of the **Water Environment Federation**

Vijay Sundaram, P.E., Stantec
  Consulting, *Chair*

Jorge T. Aguinaldo, BCEEM,
  Bios Technologies
Hardeep Anand, Miami Dade
  Water and Sewer
Lisa Andrews
Jason Assouline, P.E., CH2M
Pitiporn Asvapathanagul, Ph.D.,
  California State University,
  Long Beach
Paul Biscardi, Hazen and
  Sawyer
Thierry A. Boveri, CGFM,
  Public Resources
  Management Group
Arturo A. Burbano
Marie Burbano, CDM Smith
Daniela Castañeda, P.E., Stantec
  Consulting
Bruce L. Cooley, P.E., Black &
  Veatch
James Crook, Ph.D., P.E.
Allegra da Silva, Ph.D., P.E.,
  Stantec Consulting
Bryce Danker, Hazen and
  Sawyer
Gennaro Dicataldo, Ph.D., P.E.,
  Phillips 66 Technology
Richard Finger
Ed Greenwood, P.Eng.,
  Amec Foster Wheeler,
  Environment &
  Infrastructure
Mark Greene, Ph.D., O'Brien &
  Gere

Joseph Griffey, P.E., Smith
  Seckman Reid, Inc.
Tyler Hadacek, Stantec
  Consulting
Katie Henderson, Water
  Research Foundation
Fahim Hossain, Ph.D., Imam
  Abdulrahman Bin Faisal
  University
Bob Hultquist, P.E.
William T. Hunt, Orange
  County Water District
Stephanie Ishii, Hazen and
  Sawyer
Frank J. Johns, Tetra Tech
Kim Jusek, P. Eng., Stantec
  Consulting
Stephen Katz, GE Water &
  Process Technologies
Wendell Khunjar, Hazen and
  Sawyer
Mark Knight, GE Water &
  Process Technologies
Kevin S. Leung, Washington
  State Department of
  Ecology
Louis Charles Lefebvre,
  Cloacina, LLC
Robert V. Menegotto,
  MANTECH
John T. Morris, P.E., BCEE,
  Hon.D.WRE, WEF Fellow,
  F.ASCE, Metropolitan
  Water District of Southern
  California
Emery Myers, Stantec
  Consulting

Tyler Nading, P.E., CH2M
Smiti Nepal, U.S. EPA
Robert J. Ori, Public Resources
    Management Group
George Patrick
Patrick Regan
Carlos Reyes, Las Virgenes
    MWD
Craig L. Riley, P.E., Washington
    State Department of Health,
    Retired
Caroline Russell, Ph.D., P.E.,
    BCEE, Carollo Engineers,
    Inc.
Larry Schimmoller, P.E., CH2M
Mia Smith, Stantec Consulting
Nicholas T. Smith, Public
    Resources Management
    Group
Sheryl Smith, CDM Smith
Christopher Stacklin, P.E.,
    Orange County Sanitation
    District

Emily Stahl, City of Guelph,
    Ontario, Canada
Ben Stanford, American Water
Eva Steinle-Darling, Ph.D., P.E.,
    Carollo Engineers, Inc.
Neil Stewart, Stantec
    Consulting
Patricia Tennyson, Katz &
    Associates
Jason Turgeon, U.S. EPA
Milind V. Wablé, Ph.D., P.E.,
    BCEE
Tom Watson, City of Santa
    Monica
Brian Wheeler
A. Brooke Wright, Stantec
    Consulting
Phil Yi, Hazen and Sawyer
Usama Zaher, Ph.D., P.E.,
    INTERA, Inc.

Under the Direction of the **Water Reuse Subcommittee** of the **Technical Practice Committee**

2018

Water Environment Federation
601 Wythe Street
Alexandria, VA 22314-1994 USA
http://www.wef.org

Special Publications of the Water Environment Federation

The WEF Technical Practice Committee (formerly the Committee on Sewage and Industrial Wastes Practice of the Federation of Sewage and Industrial Wastes Associations) was created by the Federation Board of Control on October 11, 1941. The primary function of the Committee is to originate and produce, through appropriate subcommittees, special publications dealing with technical aspects of the broad interests of the Federation. These publications are intended to provide background information through a review of technical practices and detailed procedures that research and experience have shown to be functional and practical.

Contents

Chapter 2   Development of Purpose and Needs Statement
*Marie Burbano, Tyler Hadacek, Sheryl Smith, and
Hardeep Anand*

Chapter 3   **Strategic Planning and Concept Development**
*Allegra da Silva, Ph.D., P.E.; Joseph Griffey, P.E.;
Kim Jusek, P. Eng.; Carlos Reyes; Emery Myers;
Mia Smith; Emily Stahl; Neil Stewart; and
Tom Watson*

Index

List of Tables

List of Figures

Executive Summary

Engineered water reuse systems essentially harness natural cycles to augment water demands. *All* water is reused. As Pure Water Brew (www.purewaterbrew .org) states, "No matter where it comes from, we should be judging our H_2O by its quality, not by its history". Our demands for water for both human activity and the environment are now pushed to the limits of naturally available water resources in many locations. *The Water Reuse Roadmap* (referred to as the *Roadmap*) provides safe, reliable shortcuts within the natural water cycle, harnessing water where it is needed and treating it to appropriate levels for use, depending on the application.

The purpose of the *Roadmap* is to provide concise guidance on how to holistically evaluate reuse opportunities and benefits and implement or expand a program to reuse the most underutilized water resource— wastewater. The *Roadmap* begins with initial concept development; then describes stakeholder engagement, regulatory, risk assessment, and planning processes; and finishes with advice on financing, implementing, operating, and maintaining water reuse infrastructure. This is a practical water reuse planning and project implementation guide, not a design manual. As such, primary audiences for this guide are water resource planners and managers serving municipalities, utilities, industries, and the commercial sector (i.e., the people responsible for and managing the water resources available to a specific community and/or enterprise).

The *Roadmap* is designed to help water managers achieve their goals by (1) determining the social, technical, and financial feasibility of water reuse options in their specific situation; (2) initiating a water reuse program when appropriate; or (3) expanding an existing reuse program based on new approaches and opportunities for innovation. Initiating a water reuse program or expanding an existing program into new areas of reuse can be

1

a daunting task. The *Roadmap* provides detailed guidance, including case studies to help water managers holistically identify, evaluate, and implement their water reuse options. With almost all reuse systems, the beneficiaries are people and the environment.

The challenges, opportunities, and benefits of water reuse are broad and diverse. A successful water reuse program is multifaceted and engages elements of strategic planning, decision-making, technology, communications, financing, monitoring, and maintenance management. Although technologies and regulations provide solutions and pathways, ultimately the community selects and implements the appropriate water reuse solution. Therefore, clearly communicating the value and long-term benefits of water reuse—including the safeguards, technologies, and management approaches—is essential to any successful water reuse program.

Water reuse projects provide a reliable new water source, enhance resiliency and water security, control seawater intrusion and land subsidence, alleviate water stress, help us in preserving pristine water sources in their natural state, and much more. Conventional thinking has been to collect and treat wastewater to acceptable standards and move it quickly downstream via effluent disposal methods. The new paradigm is to recover, manage, and generate value from wastewater resources (water, nutrients, and others) and not waste it via traditional disposal methods that are becoming increasingly expensive. Even today, without considering long-term holistic benefits, water reuse projects may have a lower cost than wastewater disposal projects. As an example, rather than implementing expensive wastewater nutrient removal processes to treat and dispose the wastewater, redirecting it to agricultural use (where water and nutrients are both needed) may have lower capital and life cycle costs.

The focus of the *Roadmap* is beneficial reuse of wastewater from communities, industries, and the commercial sector. The *Roadmap* intentionally does not address other potential water resource-enhancing projects such as stormwater capture and reuse, on-site residential water reuse systems, and in-facility industrial water reuse projects.

The *Roadmap* provides water managers with the following:

- An overview of virtually all water reuse opportunities and issues faced by water reuse projects;
- An understanding of the elements of a water reuse project—their importance, attributes, and integration to the community, resulting in a successful water reuse program; and
- Assistance with implementing or expanding a water reuse program.

The *Roadmap* is organized into nine chapters, each focusing on the specific elements that facilitate a successful water reuse program, from concept development through implementation and maintenance.

Chapter 1—Introduction provides a general introduction to water reuse, why it may be needed, the spectrum of water reuse opportunities supported by case studies, and concepts related to integrated water resource management.

Chapter 2—Development of Purpose and Needs Statement summarizes the development of the purpose needs statement for a water reuse project. It begins with an identification of the main drivers for water reuse, with a discussion of the vision for the community and creating an initial plan. Next, the concept of "one water" approach to resource management is described in detail, including evaluating opportunities as related to water reuse with the one water approach and identifying challenges. Potential partnerships for water reuse projects are identified along with case studies and funding concepts.

Chapter 3—Strategic Planning and Concept Development lays out the conceptual elements involved in planning a water reuse project. These elements allow water reuse planning to be transparent, inclusive, and adaptive to changing conditions. Specific elements include defining the vision, goals, and landscape of the project. In other words, how to select and plan the right project for implementation involving multiple stakeholders and how to articulate the plan to community members or other project stakeholders are addressed. The chapter includes three case studies illustrating how water reuse projects were evaluated as part of an integrated water resources management approach.

Chapter 4—Regulations and Risk Assessment provides guidelines for identifying and addressing regulatory requirements and identifying and managing risk in water reuse projects. The chapter presents an overview of water rights, water quality, and application of the Hazard Analysis and Critical Control Points approach to managing risk in water reuse applications. General regulatory and risk management considerations are outlined for agricultural reuse, general urban reuse (such as landscape irrigation and in-building uses), industrial reuse, and potable reuse.

Chapter 5—Financial Sustainability lays out financial sustainability elements of water reuse projects. There are a variety of reasons why a decision-maker may be considering a water reuse program. These reasons range from water resource resiliency to environmental sustainability. However, the bottom line for water reuse projects is the method of cost recovery to implement and sustain the project. Chapter 5 presents a comprehensive overview of financial sustainability by defining what it is, best management practices to promote it, and long-range planning to achieve financial sustainability. Ultimately, the chapter seeks to help define what is the water reuse value proposition.

Chapter 6—Communication and Outreach lays out the public outreach and communications elements of water reuse projects. Water has been used and reused since the beginning of time, but public understanding and awareness of this fact is not great. Public support varies for nonpotable water reuse applications, as well as for potable reuse. Robust public outreach and communication programs are needed to inform community members about the potential benefits of water reuse projects—both nonpotable and potable—to ensure that these projects are judged on a level playing field against other water supply projects. A successful community outreach and communication program is based on trust, transparency, and consistent/ sustained outreach efforts. The project purpose and need must be clear and easy to understand. For a water reuse project to become a reality, careful strategic planning is needed, both from an engineering perspective as well as from a communications standpoint. This chapter provides guidance on how to communicate water reuse to the public and cites references for those who wish to know more.

Chapter 7—Implementation: Treatment Technologies and Other Project Elements reviews the implementation of water reuse projects, specifically focusing on treatment technologies that can reliably meet water quality criteria requirements and achieve the identified treatment goals for nonpotable and potable reuse projects. Additional technical topics relevant to project implementation are discussed in this chapter, including treatment process sizing, operability, staffing, blending, and source control, among others.

Chapter 8—Monitoring and Control summarizes monitoring and control elements of water reuse projects. In both wastewater treatment and potable water treatment, a system of monitors and controls is necessary to reliably treat water to acceptable standards protecting public health, beneficial use, and the environment. Water reuse standards and treatment goals vary for water reuse by application type, subsequent use, human contact potential, and the likelihood of affecting the environment. To meet standards, a series of treatment processes, or barriers, is required to ensure that chemical and biological constituents are removed to safe levels. As such, monitors are an important aspect of public health and environmental protection because they are used to verify treatment process integrity and help operations teams mitigate treatment failure. Monitoring data are made actionable through controls and can be further integrated through a critical control point strategy. In this chapter, monitoring and control requirements are differentiated between nonpotable reuse, indirect potable reuse, and direct potable reuse projects.

Chapter 9—Ongoing Maintenance and Monitoring Progress lays out maintenance management elements of water reuse projects. Maintenance management is vital for stewardship of the finite resources entrusted to governing agencies. Operation and maintenance of water reuse facilities can represent a significant portion of annual budgets. However, close monitoring of the systems within these facilities can reduce these costs to deliver the best possible rate to customers. The chapter provides guidance for how to integrate water reuse facilities to the community and routinely consider other non-monetarized factors to inform community engagement, public trust, and confidence.

Throughout the *Roadmap*, readers are provided with best available references on the topics associated with implementing or expanding a water reuse program. The *Roadmap* was prepared to be the concise, "one-book" reference for water reuse planning and implementation for the foreseeable future.

1

Introduction

Vijay Sundaram, P.E., and James Crook, Ph.D., P.E.

1.0   WATER REUSE AND ITS DRIVERS

Water is essential to life, and it is becoming increasingly difficult to provide adequate water supplies to an ever-increasing human population. People are adversely affected by the lack of an adequate water supply. The most acute problems are inadequate water to drink and grow food crops, both of which are essential to life. Further effects include limits on human activities and quality of life, and limits on commercial/industrial activity and community development. Water, itself, is plentiful on earth, covering 71% of its surface; however, only 0.5% of the world's water resources are available to provide for the freshwater needs of our planet's ecosystem and population. The scarcity problem is a result of the amount of water of adequate quantity and quality available in a locale relative to human needs in that locale. As urban planners, agriculturalists, industrialists, and others look to the future, they ask the question, "Where's the water?". At least a portion of the answer, in many situations, is the reuse of treated municipal wastewater for nonpotable and/or potable uses. The use of recycled water has been shown to be safe, cost-effective, and sustainable.

Drivers for water reuse projects include the following:

- Water stress (need for water),
- Conservation of existing potable supplies (reduction of fresh water use),
- Pollution abatement,
- Lack of a reliable water supply alternative,
- Need for a cost-effective new water supply alternative, and
- Environmental restoration.

2.0   THE WATER REUSE OPPORTUNITY

Water reuse is not new. We have been reusing water since the beginning of time, we just do not think of it that way. The same water molecules consumed by prehistoric species are largely still the same molecules with us today. It is the contaminants that are introduced to the water through various types of use that need to be reduced to acceptable levels before the water can be used for beneficial purposes. Virtually all uses of water, natural or by man, add contaminants to water.

The breakthrough in modern water reuse thinking and opportunities is that now we, too, can mimic or duplicate nature's processes of removing contaminants from water. Some water might not even need much treatment after being used once, depending on its quality and its planned subsequent beneficial use.

Customizing water reuse project approaches and solutions based on how the treated water will be used (i.e., its end use) is a prevalent/common strategy in the development of multiple reuse solutions. This is termed "fit-for-purpose" reuse solutions. Water reuse is currently being considered as a potential solution for various water management needs throughout the world. The challenge has been in the implementation of water reuse solutions that are robust, cost-effective, and safe with regard to public health and the environment.

3.0   DEFINITIONS

Recycled water is municipal wastewater that has been treated to meet specific water quality criteria with the intent of being used for beneficial purposes. The terms, *recycled water* and *reclaimed water*, have the same meaning and are often used interchangeably, depending on geographic location. *Municipal wastewater* is domestic wastewater that may include commercial and industrial wastewater.

3.1   Uses and Delivery Methods

These are numerous technical terms used to explain the different uses and delivery methods of recycled water. These include the following:

- *Augmentation* is the process of adding recycled water to an existing raw water supply (e.g., a reservoir, lake, river, wetland, or groundwater basin);

- *Beneficial reuse* is the use of recycled water for purposes that contribute to the water needs of the economy and/or environment of a community;

- *Environmental buffer* refers to a groundwater aquifer or surface water reservoir, lake, or river to which recycled water is introduced before being withdrawn for normal drinking water treatment before potable reuse. In some cases, environmental buffers allow for (1) response time in the event that the recycled water does not meet specifications and (2) time for natural processes to affect water quality. Where tertiary effluent is applied by spreading for groundwater recharge, the environmental buffer provides both treatment and storage;

- *Groundwater recharge* occurs naturally as part of the water cycle and/or is enhanced by using constructed facilities to add water (such as recycled water) into a groundwater basin;

- *Potable water* is drinking water that meets or exceeds state and federal drinking water standards; and
- *Dual water system* refers to two separate distribution systems, one that supplies potable water through one distribution network and one that supplies nonpotable (recycled) water through another within the same service area.

3.2 Water Reuse Options

There are the many technical terms, often used interchangeably, to explain the different options for water reuse that a community could choose from. These include the following:

- *Potable reuse* refers to recycled water that has received sufficient treatment to meet or exceed federal and state drinking water standards and is used for potable purposes;
- *Nonpotable reuse* refers to the use of recycled water that is used for nonpotable purposes such as irrigation, industrial applications, toilet and urinal flushing, and so on;
- *De facto potable reuse* is the downstream use of surface water as a source of drinking water that is subject to upstream wastewater discharges (also referred to as *unplanned potable reuse*);
- *Planned potable reuse* is an intentional project to reclaim water for drinking water. It is sometimes further defined as either direct or indirect potable reuse (IPR). It commonly involves a more formal regulatory approval process and public consultation program than is observed with de facto or unacknowledged reuse;
- *Indirect potable reuse* is the introduction of advanced treated water to an environmental buffer such as a groundwater aquifer or surface waterbody followed by normal drinking water treatment. Indirect potable reuse can also be accomplished with tertiary effluent when applied by surface spreading (i.e., groundwater recharge) to take advantage of soil aquifer treatment; and
- *Direct potable reuse* (DPR) is the introduction of advanced treated water directly to a potable water supply distribution system downstream of a water treatment facility or to the source water supply immediately upstream of a water treatment facility.

3.3 Water Types and Quality

The following terms are often used to define the types of water or different qualities of water:

- *Advanced water treatment* is a general term used to describe the overall process and procedures involved in the treatment of wastewater beyond secondary treatment to produce advanced treated water;
- An *advanced water resource recovery facility* (WRRF) is a WRRF at which advanced treated water is produced. The specific combination of treatment technologies used will depend on the quality of the treated wastewater and the type of potable reuse (i.e., IPR or DPR);
- *Advanced treated water* is water produced from an advanced treatment facility for DPR and IPR applications that augments drinking water supplies;
- *Graywater* is the term used to describe water segregated from a domestic wastewater collection system and reused on-site. This water can come from a variety of sources such as showers, bathtubs, washing machines, and bathroom sinks. Water from toilets or wash water from diapers is not considered to be graywater. Kitchen sink water is not considered graywater in many states. Many buildings or individual dwellings have systems that capture, treat, and distribute graywater for irrigation or other nonpotable uses;
- *Raw water* is surface or groundwater that has not gone through an approved water treatment process;
- *Domestic wastewater* is used water sourced from washing our food, dishes, clothes and bodies, and toilet flushing; and
- *Industrial wastewater* and *commercial wastewater* are the liquid wastes generated by industries, small businesses, and commercial enterprises and can be discharged to a sewer upon approval of a regulating authority. Some industrial wastewater may require pretreatment before it can be discharged into the sewer system, whereas other industrial and commercial wastewaters are explicitly excluded. Controlling the release of harmful chemicals into the wastewater collection system is known as *source control*.

3.4 Treatment Technology

There are terms used to describe the different types of water treatment technology that can be used to create recycled water. Some of the most common terms are defined as follows:

- *Advanced oxidation* is one of the processes that can be used as a treatment process in advanced water treatment systems. One example of an advanced oxidation process is the use of hydrogen peroxide (H_2O_2) and UV light in combination to form a powerful oxidant that provides further disinfection of the water and breaks down some health-significant chemical constituents;

- *Dual-media filtration* is a filtration method that uses two different types of filter media, typically sand and finely granulated anthracite;

- *Granular activated carbon* (GAC) is used to remove chemicals that are dissolved in the used water by adsorption;

- *Biological activated carbon* (BAC) filtration is a biofiltration process that uses GAC as support material for the microbial population metabolizing biodegradable constituents;

- *Biologically active filtration* is a biofiltration process that uses various types of support material such as anthracite, GAC, and others for the microbial population metabolizing biodegradable constituents;

- *Microfiltration/ultrafiltration* is a type of physical filtration process where wastewater is passed through a membrane to separate micro-organisms and suspended particles from the water that are larger than the filter's pore size. The typical nominal pore size of a microfiltration membrane is about 0.1 to 0.2 µm. The typical nominal pore size of an ultrafiltration membrane is about 0.001 to 0.1 µmi;

- *Multibarrier concept* refers to the use of several different treatment processes operated in series to provide redundancy and robustness in the removal of both pathogens and unwanted chemicals and to ensure that the failure of a single process does not render the system vulnerable to penetration by microbial or chemical contaminants that pose a significant risk to public health;

- *Ozonation* is the process of applying ozone (O_3) for the disinfection of water and contaminant oxidation. Ozone is a strong oxidant;

- *Reverse osmosis* is a method of removing dissolved salts and other constituents from water. Pressure is used to force the water through a semipermeable membrane that transmits the water, but stops most dissolved materials from passing through the membrane;

- *Soil aquifer treatment* occurs when water, including recycled water, percolates into the ground and is treated by physical, chemical, and biological processes that naturally occur in soil;

- *Primary treatment* is a process in which solid matter is removed from the water by sedimentation and grease and oils are removed from the water via skimming from the surface of sedimentation tanks. The remaining liquid may be discharged or subjected to further treatment;

- *Secondary treatment* is a process in which dissolved and suspended biological matter are removed from the wastewater using biological processes; and

- *Tertiary treatment*, or *advanced water treatment*, typically refers to further improvement in wastewater quality beyond secondary treatment by chemical and/or filtration processes, followed by disinfection.

4.0   THE WATER REUSE SPECTRUM

The possibilities for water reuse are a continuum, much like a spectrum of light. Within the water reuse spectrum, the varying uses of recycled water represent a large determinant of the degree of treatment needed and the necessary reliability of that treatment. A generalized visualization of this water reuse spectrum is presented in Figure 1.1. Flood irrigation of fodder crops, where there is no public contact with the recycled water, requires the least amount of treatment and reliability features to protect public health and the environment. From this baseline, the level of treatment and reliability increase as public contact and environmental sensitivity increase, and DPR requires the highest level of municipal wastewater treatment and reliability.

4.1   Environmental Restoration — Tres Rios Wetland Restoration, Phoenix, Arizona

The City of Phoenix, Arizona, constructed the Tres Rios Environmental Restoration project, which uses treated effluent from city's 91st Avenue WRRF to restore approximately 202 ha (500 ac) of vital wetland and riparian habitat in the Salt River bottomlands (City of Phoenix, 2017). This project meets Arizona Department of Environmental Quality water quality standards and creates a mutual relationship between the restored wetlands and the nearby WRRF. It also serves as a public recreation area. The City of Phoenix has observed that the lush and scenic Tres Rios wetlands are now home to more than 150 different species of birds and animals. After

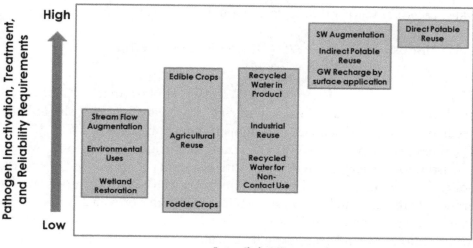

FIGURE 1.1   Water reuse spectrum.

passing through the wetlands, the remaining water is used again to irrigate crops. This project is restoring natural habitat to early 1800s conditions.

4.2 Agriculture Reuse—Monterey Salinas Valley Edible Crop Irrigation, Salinas Valley, California

The Monterey Regional Water Pollution Control Agency (MRWPCA) in the Salinas Valley of California began water reuse facilities' planning in the 1970s. The Salinas Valley is an agricultural region in northern Monterey County, where a wide variety of high-value vegetable crops are grown. A 7-year U.S. Environmental Protection Agency (U.S. EPA)-sponsored agricultural reuse demonstration study conducted at Castroville, California, determined that disinfected, filtered, secondary effluent was acceptable for spray irrigation of food crops eaten raw. The MRWPCA completed construction of a 1.3-m^3/s (30-mgd) WRRF in 1997, and began delivering 0.88 m^3/s (20 mgd) of recycled water for food crop irrigation in 1998 (MRWPCA, 2017). Recycled water is used to irrigate several types of food crops, including lettuce, celery, broccoli, cauliflower, artichokes, and strawberries. The service area is approximately 4856 ha (12 000 ac).

4.3 Landscape Irrigation—St. Petersburg, Florida

Recycled water has been used extensively for golf course and residential/commercial landscape irrigation in the Southwest and Florida. St. Petersburg, Florida, implemented one of the first significant dual water distribution systems in the United States in 1977 for various uses of recycled water, including irrigation of parks, playgrounds, school yards, and residential property. A dual distribution system is one in which pipes carrying recycled water are separate from those carrying potable water.

4.4 Industrial Reuse—West Basin's Designer Water, Los Angeles, California

West Basin Municipal Water District (WBMWD) in Los Angeles, California, produces "designer water" from municipal wastewater to meet the unique needs of various commercial and industrial customers (WBMWD, 2017). The types of designer water WBMWD produces include cooling tower water consisting of tertiary treated water with ammonia removal, low-pressure boiler feed water provided by microfiltration and reverse osmosis membranes of effluent, and high-pressure boiler feed water provided by microfiltration and two-stage reverse osmosis. The WBMWD also produces effluent for irrigation uses and groundwater replenishment, with the latter function providing the dual benefits of increased groundwater supply and reduced seawater intrusion to the aquifer. The designer water strategy enables use of recycled water far beyond the potential of common irrigation applications. This helps sustain economic production in the area by supplying recycled water to several local refineries and a power generation company.

4.5   Groundwater Recharge—El Paso, Texas

To address the depletion of the Hueco Bolson freshwater aquifer, El Paso Water in 1985 completed construction of the Fred Hervey Water Reclamation Plant (FHWRP). The FHWRP produces highly treated water using ozone and BAC advanced treatment processes. Treated water is injected to El Paso's potable water supply aquifer via 10 injection wells. The treated water meets both U.S. EPA primary drinking water standards and Texas Commission on Environmental Quality standards before injection (El Paso Water, 2007).

4.6   Seawater Intrusion Barrier and Indirect Potable Reuse—Orange County Water District Groundwater Replenishment System

Orange County Water District began its groundwater recharge efforts in the 1960s. An AWTF known as "Water Factory 21" was constructed in the mid-1970s. Treated effluent from Water Factory 21 was injected to underground aquifers to maintain a hydraulic gradient to prevent seawater intrusion. It also was an IPR project, as some of the injected water flowed inland to drinking water production wells. In the 1990s, a recharge project called the *Groundwater Replenishment System* (GWRS) was conceived. The GWRS uses reverse osmosis and an advanced oxidation process to produce highly treated water. The project began operation in 2008. The GWRS is currently producing 4.38 m^3/s (100 mgd) of water for IPR. The GWRS is planning to increase its capacity to 5.7 m^3/s (130 mgd) (OCWD, 2017).

4.7   Surface Water Augmentation—Upper Occoquan Service Authority, Virginia

In the 1970s, the Virginia Water Control Board and the Virginia Department of Health adopted the Occoquan Policy, which mandated the creation of a regional agency, the Upper Occoquan Sewage Authority (UOSA) (since renamed Upper Occoquan Service Authority), to provide state-of-the-art treatment for all wastewater generated in the Occoquan Watershed. In 1978, the UOSA Regional WRRF commenced operations with its effluent being discharged to Occoquan Reservoir, which is a significant potable water supply reservoir for the Washington, D.C., metropolitan area. This has been a highly successful water reuse project for more than 30 years (UOSA, 2017).

4.8   Direct Potable Reuse

In DPR projects, water produced by an AWTF is introduced either to the raw water supply immediately upstream of a drinking water treatment facility (DWTF) or directly to a drinking water supply distribution system. Direct discharge into the potable water supply distribution system can occur downstream of the DWTF or at some other point within the distribution system. The Big Spring, Texas, DPR project delivers advanced treated water into the raw water supply of the DWTF.

In 2016, an independent expert panel in California found that it is feasible to develop and implement a uniform set of water recycling criteria for DPR that would incorporate a level of public health protection that is as good as, or better than, what is currently provided in California by conventional (i.e., non-reuse) drinking water supplies (SWRCB, 2016). Based on the expert panel's report and ongoing research projects evaluating DPR at several locations in California, the State Water Resources Control Board currently is evaluating the feasibility of developing DPR regulations in California.

5.0   INTEGRATED RESOURCE MANAGEMENT AND WATER REUSE

Increasing water demand, uncertain weather patterns, impaired freshwater sources, and dwindling water supplies are requiring cities and industries to consider integrated resource management (IRM). Water reuse is an element of a diverse and resilient water management strategy. Robust technological solutions enable full reuse of water. However, maximum use of water through water reuse requires effective management of contaminants such as salinity, heavy metals, and chemicals of emerging concern such as hormones, pharmaceuticals, and personal care products. Water reuse solutions provide a reliable source of supply that is not weather-dependent compared to rainwater-based supplies such as river flow, reservoir volume, and, ultimately, groundwater basins. Incorporating water reuse solutions to IRM efforts is one of the tools that can be used to maintain sustainable water management in both urban and rural areas.

Integrated resource management and water reuse benefits include the following:

- Portfolio diversification. Both public and private sectors can diversify their water resource portfolios by implementing water reuse projects. Trends of extreme weather patterns, impairment of freshwater sources, and competing interests for limited water resources drive the need for a resilient and diversified water portfolio.

- Land subsidence and seawater intrusion barrier. Water reuse via groundwater recharge can be a solution for land subsidence and seawater intrusion. Damage from land subsidence (i.e., where over-drafting groundwater aquifers results in compaction of soil in the groundwater basin soil structure because of loss of void space, thus resulting in subsidence at the ground surface), coupled with rises in sea level, are projected to be significant. Implementation of ground-water replenishment projects can help to mitigate those effects.

- Alleviating competing interests for water. In areas where a limited water resource must be shared by competing interests such as municipal, agricultural, commercial, and industrial, water reuse increases the effective water resource and thereby allows increases in municipal, agricultural, commercial, and/or industrial activity/prosperity.

- Sustaining industrial activities and growth. Industries that use substantial amounts of water for cooling and/or manufacturing can collaborate with local municipalities and other stakeholders in developing projects related to integrated water management.

- Enhancing water security. Water reuse projects increase water availability and, therefore, water security.

- Environmental restoration. Environmental restoration reuse projects provide environmental benefits as well as a community's reliable water supply.

6.0   *THE WATER REUSE ROADMAP* DEVELOPMENT AND ITS ORGANIZATION

6.1   Water Environment Federation, Water Environment & Reuse Foundation, National Water Research Institute, and WateReuse Association Experts' Meeting

In February 2016, the Water Environment Federation, the Water Environment & Reuse Foundation, the National Water Research Institute, and the WateReuse Association sponsored a 2-day water reuse experts' meeting at Orange County Sanitation District (OCSD) in California to outline the next steps necessary to develop a general water reuse guide, a "roadmap", to water reuse projects throughout the United States and beyond. The OCSD experts' meeting brought together more than three-dozen representatives from water utilities, regulatory agencies, academia, consulting groups, associations, and nongovernmental organizations. These attendees built the framework for this reuse roadmap based on the following goal: "Water reuse is an element of a diverse and resilient water management strategy". Achieving this goal requires dedication to IRM to overcome public relations challenges, technical barriers, and financial constraints currently hampering water reuse, while ensuring protection of public health and the environment.

6.2   *The Water Reuse Roadmap* Primer Development

The reuse experts' meeting resulted in the development of 12 implementation elements applicable to most water reuse projects, as presented in Tables 1.1 through 1.12 (WEF, 2016). These elements are discussed in greater detail in this publication.

Each of the 12 implementation elements must be addressed to varying degrees with each water reuse project. The foregoing matrices in Tables 1.1 through 1.12 provide a synopsis of planning, implementation, and evaluation considerations for proponents of water reuse projects; moreover, they formed the basis for the development of *The Water Reuse Roadmap* (hereinafter, the *Roadmap*).

6.3   *The Water Reuse Roadmap* Organization

The *Roadmap* provides an overview of virtually all water reuse opportunities and issues for water planners, users, and suppliers considering the possibility of a water reuse project. The *Roadmap* provides directions to get a utility from its current location on the water management maturity model to where it wants to go. The *Roadmap* includes guidance on what to look for and what to avoid and ideas on some of the costs and benefits of water reuse. The book is organized into chapters under these three general categories: planning, implementation, and monitoring and maintenance.

6.4   Planning

Successful reuse projects and programs have demonstrated the value of well-thought-out planning. Planning makes proponents of reuse projects think proactively rather than reactively. Methodical planning allows for opportunities to develop and incorporate well-thought-out solutions and innovative implementation methods. Developing communication and outreach strategies and messages will be a main driver in the planning stage. Water is a very psychological subject. Therefore, strategies involving management of water, particularly reuse of water, require careful evaluation on how they affect users and/or the environment. These evaluations should drive the reuse project and its implementation; therefore, they must be

TABLE 1.1 Product development.

	PLAN	PREPARE & IMPLEMENT	EVALUATE & IMPROVE

Marketing	*Data Collection*
	• Collect data to understand the market
	  • Identify stakeholders and potential customers in the market for water reuse and other recovered resources
	  • Assess competition for product (alternative sources of supply)
	• Identify anchor customers and distribution system
	• Determine end uses, quality requirements, and variability (seasonal, daily) for customers or end use:
	  • Agriculture
	  • Environment/Habitat
	  • IPR
	  • DPR
	  • Industrial uses (cooling, washing, etc.)
		Develop Marketing Strategy
		• Create a value-cost proposition
		  • Identify value of various water quality levels for product
		  • Evaluate sales potential with respect to treatment, monitoring, and distribution costs
		• Develop marketing, sales, and branding strategy
		• Communicate benefits and advantages of water reuse
			Sell Recovered Resources
			• Brand and sell recycled water and other recovered resources
			• View customers as partners in efforts to meet customer needs through sustainable water management

Fit for Purpose	*Identify Level of Treatment*
	• Determine levels of treatment required to meet needs of customers and environment
	  • Is reuse desirable, given source water constraints, environmental needs?
	  • Is nutrient removal necessary or are nutrients desirable for irrigation?
	  • What levels of constituents (metals, total dissolved solids, etc.) are needed by end use?
	  • What are the seasonal, diurnal, or daily variations in source water and product water demands?
		Identify Opportunities
		• Produce products for market needs, such as:
		  • Irrigation (agriculture, municipal, residential)
		  • Cooling
		  • Fire protection
		  • Boiler makeup
		  • Wash water
		• Dual treatment or flexible treatment options available to produce varying quality of water for intended purposes
		• Consider decentralized infrastructure to optimize recovery
		  • Treatment (sewer scalping)
			Prioritize and Implement
			• Technical solutions design/tailored to fit-for-purpose water demands
			• Water allocation are prioritized among various users based on water supply needs and business considerations
			• Designer water approach provides operational flexibility to supply "flavors" of water

TABLE 1.2 Implementing treatment technologies.

	PLAN	PREPARE & IMPLEMENT	EVALUATE & IMPROVE

Technology Evaluation	*Identify Treatment Levels*
	• Determine level of treatment available
	• Determine level of treatment required or desired
	• Define operational/process changes required to provide water quality
	• Identify available technologies to provide appropriate multi-barrier protection.
	  • Where used/experience with technology
	  • Maturity of technology
	  • Alternative analysis
	  • Waste stream (brine, other) implications
	  • Regulatory issues
		Identify Opportunities
		• Ensure adequate treatment vs overtreating to meet regulatory requirements with minimum concentrate generation
		• Consider storage
		  • Emergency
		  • Process upsets
		  • Demand variability
		  • Equalization
		  • Monitoring
		  • Attenuation
		• Identify additional opportunities requiring more time or capital to implement, and develop a plan to finance/implement
		• Assess liquid vs solid recovery (water reuse vs land application/struvite recovery)
			Evaluate and Implement
			• Multi-barrier approach using cost-effective and low carbon footprint technology to provide right quality reuse
			• Unintended consequences are evaluated through scenario planning or other means, such as
			  • No return flows
			  • Collection system issues from scalping
			  • Aggressive water
			• Identify research and development needs to drive innovations
			• Identify water quality trading and greenhouse gas offset credit opportunities

Treatment Management	*Plan for the Future*
	• Identify unit operations/basins for use in future iterations of designer water production
	• Long-term planning such as leaving space in the facility hydraulic profile to accommodate future processes
	• Develop scenario analysis in master planning
	  • Future regulations
	  • Water supply/demand
	  • Treatment resiliency and failsafe planning
		Mitigate Risks
		• Design for current requirements with an eye toward future requirements
		• Validate technologies
		  • Reliability
		  • Long-term prospects
		  • Path dependency
		  • Public support
		  • Intellectual property issues
		  • Operational efficiency
		  • Scalability
		  • Monitoring requirements
		  • Byproducts and coproducts
			Manage Tradeoffs
			• Understand tradeoffs
			  • Reliability vs advanced technology
			  • Regulatory requirements vs business needs
			  • Resource recovery vs treatment

TABLE 1.3   Monitoring and control.

	PLAN	PREPARE & IMPLEMENT	EVALUATE & IMPROVE

Monitoring	*Collect Information*
	• Parameter monitoring for decision-making, potentially includes
	  • Upstream
	    • Collection system
	    • Industrial dischargers
	  • At treatment facility
	    • Influent, in-facility, and effluent
	  • End use
	    • Customer (agriculture, industrial)
	    • Environment (receiving body, aquifer)
	• Laboratory capacity at facility and throughout region analyzed for analytical, biological services for cation-exchange capacity, CCL3, and unregulated contaminant monitoring rule
	• WRRF considers frequency of sampling to account for diurnal and seasonal variations
		Analyze Data
		• Process information to understand options, examples include
		  • Rate of accumulation and control of salts and metals
		  • Technical resiliency (e.g. dynamic response time to spike loading)
		• Model the treatment process to understand its constraints and opportunities, and ensure redundancy in design
		• Assess current facility performance for varying water quality needs of customers
		• Bioassay tools for process monitoring are developed
		• Reliable online/early warning monitoring systems are in place
		• Data analytics management plan developed
			Proactively Use Data
			• Use data to improve
			  • Facility operations
			  • Trading programs
			  • Source control
			  • Next design upgrade
			  • Continual improvement of monitoring program
			• Sensor and monitoring advances for reporting are tracked
			• Well-developed laboratory capabilities and/or partnerships for performance and compliance
			  • UCMR
			  • CCL3
			  • LEC
			• Unregulated water quality parameters

Process Control	*Get the Big Picture*
	• Baseline performance (water quality, energy use, etc.) and benchmarks are determined
	• Supervisory control and data acquisition (SCADA) and other control system capabilities and needs are identified
		Understand Key Processes
		• Proactive maintenance is in place through computerized maintenance management system
		• Technologies for remote monitoring system are in place (e.g., wide integration including satellite systems)
		• Real-time monitoring and control strategy in place
		• Developed mass balances
		  • Water
		  • Organics (energy)
		  • Nutrients (nitrogen and phosphorus)
		  • Salts
		  • Metals
			Monitor for Real-Time Control and Optimization
			• Real-time control is in place (e.g., SCADA) to optimize water quality, chemical use, energy use, and efficiency
			• System learning algorithms (data driven) that incorporate dynamic supply–demand challenges are in place

Quality Control	*Product Development* • Develop a quality assurance program and process for products • Implement adaptive management techniques	*Product Production* • Quality assurance processes in place • Water quality (including chemical addition) • Water quantity (to avoid supply disruptions)	*Quality Verification* • Utility has adopted a quality standard (HACCP, Six Sigma, ISO 9001) • Permit and compliance program exists with established water analysis protocol • Active partnership with agriculture exists to ensure downstream product quality

TABLE 1.4 Implementing innovation.

	PLAN	PREPARE & IMPLEMENT	EVALUATE & IMPROVE
Research and Development (R&D)	*Prepare for R&D* • Collaboration with research organizations as WE&RF drives innovation and understanding and adoption. • Staff well-versed in existing technologies • Opportunities are identified by survey of emerging technologies • Reduce risk through collaborative research and information sharing • Have WRRF leadership/managers recognize and reward innovative approaches	*Preform R&D* • Utility budget includes R&D funding to demonstrate culture of embracing innovation • Utility actively participates in water innovation partnerships (e.g., Water Innovation Centers, research foundations, university partnerships, etc.)	*Expand R&D* • Site visits to facilities utilizing innovative technologies occur regularly • Completed trials and research projects provide the foundation for further advancement within the industry • Utility serves as a demonstration facility for public education and collaboration with research organizations like WE&RF
Test Beds	*Evaluate Technologies* • Technologies that reduce energy use or increase generation are identified • Test beds are identified to enhance collaboration with universities, R&D of the equipment supplies and agencies and other stakeholders • Validation and data requirements for new/alternative technologies are identified	*Initiate Trials* • Treatment technologies are demonstrated • Standard validation protocols developed to evaluate technologies for effectiveness and sustainability • Develop mobile treatment testing for field testing • Plans are made to leverage test beds and communicate results with data base of results • Institutional resistance is overcome by demonstration facility, pilot projects, specificity of design and technology options, and goals of the reuse project	*Implement Full-Scale Solution* • Flexible and effective technologies are implemented to meet the needs of the various quality levels of recycled waters

Alternative Management Approaches	*Identify Alternatives*
	• Decentralized treatment options are considered
	  • Package plants and technologies
	  • Scalping plants
	  • On-site reuse
	  • Green infrastructure
	• Planning is performed on a watershed basis and includes consideration of robustness of satellite nodal structure of decentralized infrastructure
		Implement Alternatives
		• Green infrastructure technologies such as treatment wetlands and riparian buffers are implemented where appropriate as part of multiple barriers approach
		• Enhanced regionalization (e.g., biosolids processing) has been considered and implemented where appropriate
			Expand Integration
			• Alternative management approaches (e.g., decentralization, regionalization, etc.) are used, where appropriate, to maximize overall, regionwide benefit

TABLE 1.5 Message development.

	PLAN	PREPARE & IMPLEMENT	EVALUATE & IMPROVE
Legitimacy	*Identify Resources* • A transparent planning process is developed to establish legitimacy of project owner • Independent advisory group or expert panel considered • Success stories and best practices from relevant organizations are identified	*Collaborative Communication* • Appraisal conducted of public perception of: utility competence legitimacy, public health issues, water supply, risks • Plans in place to address Pragmatic, Moral, and Cognitive Legitimacy	*Ongoing Leadership* • Legitimacy of project and project owner established to meet the region's water needs • Organizational commitment to public health protection is accepted by public • Utility is the trusted source of quality
Stakeholder	*Identify Stakeholder Values* • Identify values of community and utility board of trustees • Shift cultural mindset from "meeting the permit" to recovering resources • Identify environmental, social, and economic incentives for water reuse • Identify visible agency champion	*Public Outreach & Intake* • Develop public understanding of the new purpose of a WRRF • Gather input from all stakeholder categories	*Shared Experience* • Share best practices with other utilities and the sector • Identified stakeholder groups vested in each benefit/service

Message Development	*Develop Message*
	• Understand local cultural values
	• Positive messaging (NOT "do not drink from purple pipe") developed for reuse
	Themes to include
	  • Resiliency
	  • Water independence
	  • Drought proofing
	  • Sustainable
	  • Safety
	  • "All water is reused" – de facto reuse
	  • "Water should be judged by its quality, not its history"
	  • "One Water" – reuse is just another part of the portfolio
	  • Value of water
	  • Water as a valuable service, which enhances health and wealth of community
	  • Water as critical product for agriculture, industry, commercial uses and the associated economic development
		Enhance Message
		• Consistent terminology used
		• Counter-argument developed to counter false information
		• Message uses easy-to-understand common terminology for reuse technologies and presents them in exciting and enticing ways
			Continually Evaluate Message
			• All utility members know the "elevator speech" of the reuse mission and can clearly articulate the message whether at work or in the community
			• Partner agencies are consistent in messaging on reuse
			• "Flavors" of water used to demonstrate fit for purpose
			• Measures of public opinion are used on a periodic basis to evaluate program and update messaging to ensure public acceptance
			• Ongoing engagement with opponents with respect
			• Educational activities are engaging for appropriate audiences (e.g., field trips focusing on environment and technology for students, reuse beer tasting for adults)

TABLE 1.6 Communication and outreach.

	PLAN	PREPARE & IMPLEMENT	EVALUATE & IMPROVE

Customers and Community	• Community education on fit for purpose water, the urban water cycle, and is continuously updated. Education precedes customer outreach and education strategy on specific project needs to develop understanding
	• Community groups are identified for outreach to develop understanding
	• Early education and outreach to the public avoids opposition based on yuck factor for all reuse sources but particularly drinking water
	• Agency obligation to initiate discussion with the public
	• Ensure that "Community" leaders are involved in the decision process
	• Speakers Bureau trained for effective communication with the public (educated volunteers)
	• Public volunteers representing different interests are essential to serve as community ambassadors
		• Proactive customer education and outreach program (e.g., bill inserts, tours, fact sheets, website) that focuses on public health, economic growth, environmental benefits, and cost-effectiveness is established
		• Community opinion leaders are engaged
		  • Politicians
		  • Reporters
		  • Environmental interests
		  • Clergy
		  • Physicians and other medical professionals
		  • Farmers
		  • Teachers
		  • Service groups
		• Early adopters are identified and encouraged to educate those with questions (doubters)
			• Utility engages customers in helping to achieve sustainable water resources management
			• Utility and public employees are ambassadors to community
			• Transparent process for community-driven decision-making is in place to demonstrate respect for community
			• "Hands on" engagement opportunities exist
			  • Demonstration and visitor center
			  • Interactive and static displays and exhibits
			  • Water tastings
			  • Beer tasting
			  • Technology tours
			• Utilize information that already exists through the WateReuse Association and others
			  • "Downstream"
			  • "The Global Connections Map
			  • "The Ways of Water" video
			  • Water: Think & Drink animations

Media	• Media outlets are identified and strategies are developed	• Media kit is developed (e.g., video, sound-bites, pictures, and press releases) • Deal proactively with national water crisis in news (Flint, food contamination, floods, droughts, etc.)	• Dedicated utility staff work on messaging with media, both news and social • Media champions are nurtured to inform public of factual benefits and issues with respect to reuse, especially potable applications

Water Sector	• Key energy staff network at local/regional industry events and information sharing groups supporting
	  • National goal supporting reuse and conservation
	  • Uniform criteria for IPR/DPR projects
	  • Risk-based reuse guidelines for agriculture, industrial, and commercial projects
		• Successes, failures, and lessons learned are shared at industry events
		• Staff contributes to development of guidelines for design and operation of relatively newer treatment and monitoring technology
			• Staff leads industry initiatives to support sector advancements in water reuse, specifically
			  • National reuse framework
			  • National regulatory standards

TABLE 1.7 Risk management and communication.

	PLAN	PREPARE & IMPLEMENT	EVALUATE & IMPROVE

Public Health	*Identify Public Health Issues*
	• Engage public health tracking networks (Center for Disease Control and Prevention)
	• Project plans consider public health and environmental risk for both new and established riparian projects
		Mitigate Public Health Risks
		• Systems approach is used for reliability analysis mitigating risks of
		  • Contaminating potable water source IPR
		  • Recontamination of groundwater
			Enhance Public Health Network
			• Multiple barriers are in place to reduce health and environmental risks
			  • Robust/redundant treatment
			  • Cross-connection control
			• Real-time pathogen monitoring and risk management throughout treatment and distribution

Risk Management	*Identify and Prioritize Risks*
	• Risk of high consequence of a single dramatic failure is understood
	• Strategy for risk mitigation is developed
	• Planning includes consideration of
	  • Measures for climate change adaptation (e.g., extreme events)
	  • Overtreatment vs public safety
	  • Water shortage
	  • Not recovering reclaimed water in GWR/augmentation projects
	  • Treatment technology
	  • Cyber technology
	  • Unintended consequences
	  • Health, environmental, social
		Mitigate Risks
		• Risk is reduced through water source diversification
		• Risk allocation ensures each party takes the risk that they can manage/control
		• Cyber-security capabilities and contingency plans are in place
			Leverage Innovation
			• Have contingencies to deal with uncertainties
			• Technology risk level set so that no single failure of an active component shall result in health risk
			• Organization can successfully trial and implement innovative projects and is adaptable to emerging opportunities and changing environments

Risk Communication	*Define Risk Message*
	• Agency, community, regulatory risk tolerances identified
	• Risk concepts include
	  • Opportunities instead of challenges
	  • Uncertainty instead of risk
	  • Cost vs risk
	  • Risk of status quo
	  • Risk vs benefits, without overselling
	  • Testing and monitoring framework and controls in place
	  • Needs vs risk
	  • Meeting regulations vs cost of overtreating
	  • Zero risk does not exist, and safety is what the public wants
		Refine Risk Communications
		• Risk information on CECs, toxicology, and other probabilistic statements are crafted in plain English
		• Communication efforts learn from other sectors (HACCP, etc.)
			Validate Approach
			• Independent expert panel in place to evaluate and provide oversight for risk management

TABLE 1.8   Regulatory environment.

	PLAN	PREPARE & IMPLEMENT	EVALUATE & IMPROVE

Regulatory and Legislative	*Identify Regulations*
	• Key regulators are identified and effective working relationships are established prior to project development
	• Legislative strategy is developed to enhance opportunities and minimize hurdles for water reuse
	• Regulatory framework and gaps are identified
	• As part of an ongoing relationship, key regulators are educated on potable and nonpotable reuse issues
		Seek to Unify Regulations
		• Utility advocates for unified regulations that
		  • Are science based
		  • Protect health and environment while maintaining flexibility for types of technologies
		  • Incentivize collaboration across district borders and across disciplines
		  • Are flexible for small systems
		  • Encourage innovation
		  • Are clear and consistent, without conflicts (such as between National Pollution Discharge Elimination System and recycled water permits, or water vs air, groundwater vs watershed)
		• Regional collaboration with other agencies occurs (e.g., for funding or policy changes)
			Resolve Differences
			• Utility works with industry associations to influence regulators/legislature to create incentives to encourage reuse, where appropriate
			• Utility influences funding agencies to prioritize water sector projects
			• Regulators and utility work together to resolve cross-media issues
			• Mechanism exists for resolving differences in conflicting regulations, dealing with externalities, and creating the space for innovation

Regulatory Risk Management	*Identify and Prioritize Risks*
	• Evaluate legal/regulatory implications of voluntary action
	• Identify early technology adoption risks
	• State-of-the-art technology vs new technology vs technology specified by regulation
	• Evaluate costs of overtreatment or regulations that are not practicably implementable
		Mitigate Risks
		• Develop strategy for risk mitigation and/or sharing
		  • "Creating the Space for Innovation"
		  • "Safe Harbor"
		• Encourage policies that enable rapid new technology evaluation and adoption
			Leverage Innovation
			• Organization successfully implements innovative projects and is adaptable to emerging opportunities
			• Anticipate future regulations and impacts
			• Organization supports integrated, systems-thinking approach to regulations, such as
			  • Reuse as crop irrigation source and other Food Energy Water Nexus issues
			  • Public–private partnerships

Water Rights	*Evaluate Legal Framework*
	• Evaluate positive and negative implications of policies such as
	  • Conservation mandates
	  • Water rights (agriculture, urban)
		Manage Permits
		• Permitting reuse projects address legal risks of water rights
			Source Ownership
			• Ownership of source (wastewater) and the product (recycled water) is well-defined

TABLE 1.9 Local issues.

	PLAN	PREPARE & IMPLEMENT	EVALUATE & IMPROVE

Local Drivers	*Identify Drivers*
	• Key drivers are considered, including
	  • Economic development
	  • Water scarcity
	  • Climate change
	  • Regulatory requirements
	  • Urbanization
	  • Diversification of water sources
	  • Reduced dependency on external sources
		Evaluate Impacts
		• Evaluation of impacts to local needs developed
		• Project boundaries are identified
		• Economic impacts quantified (gross domestic product growth, job creation)
		• Natural resources, impacts are identified (land subsidence, wildlife habitats, water sources)
			Maintain Vigilance
			• Plans are in place to adapt to changes of drivers

Integrated Water Resources Management for "One Water"	*Evaluate Opportunities*
	• Explore and analyze opportunities for collaboration on water resources between water, wastewater, and stormwater utilities, as well as other water stakeholders such as those in industry and agriculture
	• All available sources of water are identified, including reuse, surface water, groundwater, seawater, stormwater
	• Potential partners and neighboring agencies are involved in watershed-based planning
	  • Utilities
	  • Agriculture
	  • Industry
	  • Power plants
	  • All relevant sectors
		Establish Connections
		• Utility planning efforts are integrated with other agencies regarding multiple resources (e.g., water, stormwater, etc.)
		• Implement contracts with partners to facilitate data exchange and planning
		• Regulatory issues have been addressed and satisfactorily resolved
		• Consider embedded resources in water such as nutrients and energy in water resource planning
		• Consider accumulative constituents (salts, metals) and develop long-term management strategy
		• Partners understand relationships between treatment, discharge, and reuse
		• Framework for future potable reuse developed, even if initial reuse is nonpotable
			Leverage Resources
			• Holistic evaluation methodologies (e.g., triple-bottom-line) are used to regional (watershed) water supply plan
			• Mature protects participate in dialogue to promote water valuation that is consistent with regional/watershed sustainability
			  • Return to river
			  • Local issues (desalination plant next to flood farming)
			• Optimized supply demand model exists of future growth and resiliency plans
			• Opportunities exist for water trade between community and regional users
			• Water resources managed in an integrated manner (integrated water resources utility with water/wastewater or multi-agency approach with embedded leadership at all levels of organization)

Collaborative Partnerships for "One Water"	*Evaluate Opportunities* • Identify markets/opportunities for reuse water, as well as applicable water quality and treatment requirements • Source water and treatment objectives are identified according to market and water quality needs	*Establish Connections* • Connect with customers and potential customers to ensure they understand opportunities for diverse, sustainable sources of water	*Leverage Resources* • Utility uses partnerships to maximize water reuse sales revenues and/or reduce demand for water and energy, and optimize the need for advanced and costly levels of treatment • Systems thinking integration of systems – water, energy, food – is in place

TABLE 1.10 Strategic management.

	PLAN	PREPARE & IMPLEMENT	EVALUATE & IMPROVE

Vision	*Develop Vision*
	• Leadership Group develops Water Reuse Vision
	  • Long-term 5 years to 50 years
	  • Resilience and drought-proofing
	  • Water independence
	  • Flexibility
	  • Utility financial sustainability
	  • Economic development
	  • Environmental enhancement
	  • Resource recovery
	  • Health and ecological
	  • Become climate-ready
	    • Reduce direct and indirect emissions
	    • Improve resilience – address water scarcity and diversify water sources
	    • Improve efficiency of water supply
		Communicate Internally
		• WRRF leadership/managers link the vision to staff performance plans
		• WRRF leadership/managers incorporate sustainability goals and key performance indicators into strategic plan
			Communicate Externally
			• Utility shares vision with external stakeholders and the industry
			• Long-term, yet flexible, plans are in place to embrace external market changes
			  • Future regulations
			  • Political outlooks
			  • Changing demand
			  • Changing public acceptance
			• Robust asset and risk management in place
			• Review performance against goals
			• Reassess long-term goals

Strategic Direction	*Set Goals* • Goals and key performance indicators are established for both water conservation and water reuse to encourage efficient use of integrated water resources • Investigation conducted of organizational capacity (technical, managerial, and financial) to handle interdependencies of a reuse program	*Gather Support* • Utility incorporates goals and key performance indicators into strategic plan • Adequate organizational capacity exists to ensure effective practices can be developed and adopted to meet demands of reuse programs	*Prioritize & Implement* • Program initiatives are prioritized using tools such as Strategic Business Planning • Utility utilizes triple-bottom-line approach for sustainability project decision-making • Interdisciplinary collaboration exists to ensure thorough understanding of timescale and resources required for reuse programs

Staff Development and Alignment	*Set Training Plan*
	• Utility fulfills training needs for all relevant positions: management, engineering, and operations
	• Current practices are evaluated for training needs with respect to potential impact of water reuse/resource recovery activities
	  • Technology complexities
	  • Flexibility to meet multiple objectives
		Train and Support Staff
		• Relevant staff are trained according to current and future knowledge requirements, including
		  • Technology operations
		  • Potable applications constraints
		  • Predictive analytics
		  • Safety
		• Staffing and institutional knowledge issues, including recruitment, enhancement, and succession are planned for
			Empower Staff
			• Standard operating procedures and SCADA system are suitable for new processes and technologies
			• Operations staff are certified in both wastewater and drinking water operations
			• WRRFs mentor and guide other local and regional utilities to advance reuse goals
			• Team connects to other organizations to maintain currency of knowledge

TABLE 1.11   Financial sustainability.

	PLAN	PREPARE & IMPLEMENT	EVALUATE & IMPROVE

Benefit Identification	*Identify Benefits*
	• Benefits of water available through reuse are identified and associated with specific beneficiaries
	• Benefits are quantified
	• Long-term view is undertaken to represent future generations
		Analyze Benefits
		• Benefits are expressed in monetary terms to demonstrate the inherent value in wastewater
		• Benefits are evaluated on triple-bottom-line for sustainability
		• Deferred capital and other opportunity costs related benefits are identified and quantified
			Obtain Benefits
			• Benefits are internalized through reduced costs and revenue enhancement
			• Reward incentives are created for operations staff
			  • Energy efficiency
			  • Water resource management
			  • Inorganic salt management
			• Carbon management resource recovery

Cost Identification	*Identify Costs*
	• Costs of service for reuse are identified
	  • Treatment, distribution, marketing, customer service
	  • Capital, operations and maintenance
	  • Absolute and marginal
		Allocate Costs
		• Costs are allocated to functions and organizations (water, wastewater, customer)
		• Ancillary benefits to water reuse are included in cost analysis with respect to other water sources to ensure consistent calculation and comparison between alternatives
			Ensure Fairness
			• Environmental justice/water as a basic human right is recognized and accommodated in the cost model to ensure affordability

Financial Viability	*Identify Funding Options*
	• Develop financial strategy to support water reuse projects
	• Sources of potential funding are identified
	  • State Revolving Fund (SRF), federal, and state grants
	• Consider alternate financing methods
	  • Public–private partnerships (P3) and joint ventures
	• Alternative project delivery (design-build-operate, build-own-operate-transfer, etc.)
	• Recovery of costs from ratepayers is evaluated to ensure full understanding of legal framework (including regulatory and tax reform laws)
	• Incentives and tax credits for new/innovative or available technologies are investigated
		Budget for Success
		• Use life-cycle analysis for project decision-making
		• Water reuse is considered on all capital project designs, in operating budget decisions, and standard operating practices
		• Sustainable revenue model is in place to address
		  • Equitable cost allocation
		  • Variability of demand (seasonality, drought, abundance)
		  • Diversification of financial portfolio
			Invest in the Future
			• Effective utilization of available sources of funding (public, private, bonds, SRF, grants)
			• Full value of water is recovered in reuse rates, including variable costs based on "flavor" of water used and incentives for upstream reuse and recycling
			• Local industry and NGOs are engaged to ensure continuous support for funding
			• WRRF's recovered resource revenues generate sufficient funding to invest in other priorities and reduce upward pressure on rates

TABLE 1.12 Resiliency.

	PLAN	PREPARE & IMPLEMENT	EVALUATE & IMPROVE

Resiliency	*Evaluate Water Resources*
	• Available water resources are quantified to evaluate resiliency of current supply and the current systems ability to meet future demands based on growth and supply shortage
	• Importance of diverse water supply portfolio is understood, including
	  • Reuse
	  • Surface water
	  • Groundwater
	  • Seawater
	  • Stormwater
	  • Conservation
		Implement Water Supply Plan
		• Methods developed to define value for long-term sustainability through diversification
		• Water portfolio and resiliency and sustainability
		• Climate change, urbanization, etc.
			Optimize Supply
			• No stranded assets as full utilization of reuse as a new water supply
			• Diverse water resource portfolio is resilient and adaptable to changes in demand, water quality

Continuity of Operations	*Prepare Plans*
	• Integrated continuity of operations plan is developed
	  • Natural disasters (hurricane, earthquake, flooding, fires, etc.)
	  • System integration risk
	  • Water quality dynamics impacting customer agreements
	  • Public safety
	• Worker safety (gas, chemical spills, etc.)
		Test, Implement, Improve
		• Plan is tested and reviewed for currency on annual basis
			Continuous Monitoring
			• Proactive controls and monitoring analytics manage process risk

communicated effectively. Potential risks posed by water reuse projects vary based on the type of reuse project being proposed. All potential risks must be identified, evaluated, and, where deemed necessary, mitigated to ensure public health and/or environmental protection. Risks and the associated mitigation measures must be identified and communicated to stakeholders. Water reuse solutions are very community-specific, and must be based on local factors. There is no one "cure all" water reuse project. Water reuse planning should present an analysis of overall community benefits derived from an improved water portfolio.

The following *Roadmap* chapters are related to planning:

- Chapter 2, Development of Purpose and Needs Statement,
- Chapter 3, Strategic Planning and Concept Development,
- Chapter 4, Regulations and Risk Assessment,
- Chapter 5, Financial Sustainability, and
- Chapter 6, Communication and Outreach.

6.5   Implementation

Successful implementation of a water reuse project is more than merely delivering recycled water to a point of use. Implementation should incorporate approaches that are suitable for the local conditions. Although there is no single "best path" to implementation that applies to all types of recycled water projects, some of the steps that should be taken to achieve successful implementation of a project include, but are not limited to, the following:

- Identification of project needs and drivers;
- Identification and characterization of recycled water sources;
- Public involvement program;
- Environmental issues and approval;
- Economic and financial review;
- Regulatory review and approval, including funding mechanisms;
- Construction;
- Permitting process;
- Project commissioning and delivery of services; and
- Operation and maintenance.

In general, projects that are to be successful require public and political support; sufficient demands and returns to justify investment (the community's return on investment may not always be monetary); sources and supplies that match demands (quantity, timing, quality); adequate funding/financing; and other criteria specific to the project or jurisdiction (USBR, 2004). Chapter 7, Implementation: Treatment Technologies and Other Project Elements, is the *Roadmap* chapter related to implementation.

6.6   Operation, Maintenance, and Monitoring

The success of a water reuse project is determined by its beneficial effects on the community over the years. To produce a reliable water resource that meets a community's needs, water reuse assets must be operated, maintained, and monitored effectively. Although operation, maintenance, and monitoring needs vary by the type of reuse project implemented, as implied by the water reuse spectrum presented in Figure 1, there are basic principles that apply to all water reuse projects. The operation, maintenance, and monitoring of water reuse infrastructure often requires a sizable portion of annual budgets. Proactive approaches, including upfront investment in system resiliency and innovation, are important considerations when managers evaluate options to optimize efficiency, maximize recovery of water, and meet all regulatory requirements.

The following *Roadmap* chapters are related to operation, maintenance, and monitoring:

- Chapter 8, Monitoring and Control and
- Chapter 9, Ongoing Maintenance and Monitoring Progress.

7.0   REFERENCES

City of Phoenix (2017) Tres Rios Wetland Project Webpage. https://www.phoenix.gov/waterservices/tresrios (accessed May 2017).

El Paso Water (2007) http://www.epwu.org/wastewater/fred_hervey_reclamation.html (accessed May 2017).

Monterey Regional Water Pollution Control Agency (2017) http://www.mrwpca.org/about_facilities_water_recycling.php (accessed May 2017).

Orange County Water District (2017) OCWD Groundwater Replenishment System. http://www.ocwd.com/media/4267/gwrs-technical-brochure-r.pdf (accessed May 2017).

State Water Resources Control Board (2016) http://www.waterboards.ca.gov/drinking_water/certlic/drinkingwater/requirements.shtml (accessed June 2017).

Tchobanoglous, G.; Cotruvo, J.; Crook, J.; McDonald, E.; Oliviery, A.; Salveson, A.; Trussell, R. S. (2015) *Framework for Direct Potable Reuse;* WateReuse Association: Alexandria, Virginia.

United States Bureau of Reclamation (2004) *Recycled Water Project Implementation Strategies Technical Memorandum;* Report prepared for the United States Bureau of Reclamation by CH2M Hill, Santa Ana, California.

Upper Occoquan Service Authority (2017) https://www.uosa.org/Display ContentUOSA.asp?ID=1021 (accessed May 2017).

Water Environment Federation (2016) *The Water Reuse Roadmap Primer.* http://www.wef.org/globalassets/assets-wef/direct-download-library/public/03---resources/wef_water_reuse_roadmap_primer.pdf (accessed May 2017).

West Basin Municipal Water District (2017) http://www.westbasin.org/water-supplies/recycled-water (accessed May 2017).

2

Development of Purpose and Needs Statement

Marie Burbano, Tyler Hadacek, Sheryl Smith, and Hardeep Anand

1.0 DEVELOPING THE VISION FOR WATER REUSE

Around the world, utilities face numerous water resources challenges including limited fresh water supplies, drought, flooding, water quality degradation, loss of ecosystems, and aging infrastructure. Utilities have been successful in combating these challenges by building new water works, advancing water and wastewater treatment, channeling stormwater, and restoring streams and rivers. However, limited financial and human resources, increasing environmental regulations and urban development, limited fresh water, and global climate change all put new pressures on these solutions. For many utilities, these pressures call for a one water view of the watershed and the development of water reuse projects.

1.1 Drivers for Water Reuse

The drivers for water reuse are numerous and tend to vary by geographical and local infrastructure conditions. For a semi-arid area like Southern California, utilities such as the City of Los Angeles and West Basin Water District look to recycled water as a drought-proof source that can reduce the demand on costly imported water. Recycling wastewater may be less expensive than disposal options. For areas with an abundance of rain such as Florida, competing interests for water with limitations on groundwater supplies and environmental and water quality issues, plus Everglades restoration, can add demands to the available water. Additionally, utilities are seeking to become more financially and environmentally resilient, and to be good stewards of the environment. This can be in the form of financial resiliency, where the cost of water will not be subject to drastic shifts adding stress to the community. It can also be environmental resiliency, where a utility can be prepared for drastic climate change phenomena or changes in the water supply. Businesses and industry are also looking for resiliency. Water costs and availability can affect production and, therefore, are a form of risk that needs to be managed. Overall, industry can also reach out to municipalities to form a partnership. Table 2.1 presents common drivers for water reuse.

1.2 Developing a Vision for the Community

Although utilities are typically the main drivers for a water reuse project, the vision for the project can be considered a vision for the community. In this case, the community can include the local utility or owners, regulators, elected officials, community organizations, industries that use a lot of water, private water users, and facilities. In fact, the success of a water reuse project typically corresponds to the ability of the lead agency or agencies to successfully work with all members of the community. For some utilities looking

TABLE 2.1   Drivers for water reuse.

Driver	Description
Economic development and urbanization	Recycled water can be a best-value proposition for providing a lower-cost water than alternatives
Water scarcity and resource recovery	Use wastewater as a valuable resource to be conserved and reused, especially in areas where water resources are limited
Resiliency and climate change	An additional source of water to alleviate pressure on existing drinking water systems
Regulatory requirements	Meet rules and regulations from federal, state, or local governments with increased pressure from tighter discharge regulations lead utilities to look at alternatives to discharge such as reuse
Diversification of water sources	Provide multiple sources of water in case of degradation or damage to a single source
Reduced dependency on external sources	Reduce the need for imported water or make sources available to environmental restoration
Need for financial resiliency	Limited funding and a growing financial need for best value solutions
Reduce or eliminate an existing discharge to the environment	Decrease ocean outfall, deep injection, and surface water discharges
Excess water management	Manage high flows to prevent flooding

to reuse water, there can be a disconnect because acute scarcity is not seen as a key issue. In other cases, there is a clear lack of understanding and communication between stakeholders. Having all stakeholders on the same page in terms of understanding is what can bring acceptance and success.

Recycled water projects rely on a high level of agency, regulatory, legal, and funding support, including formalized agreements between separate owners and staff buy-in for implementation. The development of a vision for the community is critical to creating alignment of the key stakeholders to the overall purpose of the project. By formulating the vision, it confirms everyone's understanding of the project drivers, goals, and scope in terms of what is included and what is not included. In addition, formulating a vision provides the opportunity to ask both "what" the project is going to accomplish and "why" it is important for each of the partners. The discussion of importance and value of the project is an effective way to strengthen the sense of commitment to the project and the pride in being a key stakeholder, community member, or organization staff. In developing a vision, the driving group should consider the following:

- Short and long-term goals,
- Resilience and drought-proofing,
- Water independence,
- Flexibility,
- Utility/owner financial sustainability,
- Economic development,
- Environmental enhancement,
- Resource recovery,
- Health and ecological,
- Potential funding sources,
- Current water use,
- Current discharges to the environment, and
- Triple-bottom-line benefits.

1.3   Creating an Initial Plan

Implementing a recycled water facility is a complex task that involves the efforts of many diverse organizations, including utilities, regulators, designers, contractors, and the community. Each phase of implementation must be successfully completed to keep the project moving toward physical construction. Table 2.2 lists key questions to consider when developing the initial plan for a recycled water project.

The initial plan should include a fatal flaw analysis early in the process. Salt management and brine disposal frequently constrain reuse options. The availability of a reuse distribution system, particularly with the use of "purple piping" in urban areas, can also be a fatal flaw.

2.0   "ONE WATER" APPROACH TO RESOURCE MANAGEMENT

2.1   Integrated Water Resources Management for "One Water"

As utilities and owners respond to the challenges of limited fresh water supplies, drought, flooding, water quality degradation, loss of ecosystems, and aging infrastructure, there is a shift to a more holistic approach to water

TABLE 2.2   Initial planning for a recycled water project.

Category	Challenges
Management	• What are the local community needs and potential effects? • Does the project have adequate agency support? • What kind of and how many agreements are necessary? • Who will own the project: a water agency or a water reclamation agency? • What kind of organizational effects will there be? • What are the risks? • What are the unknowns? • What treatment standards apply? • What are acceptable levels of risk as the owner? • What other organizations should be involved as a partner? • Which stakeholders should we include?
Design	• What needs to be included in the preliminary design to define the project? • What types of construction procurement are best? • Is pilot testing required and for what/how long? • What level of owner participation is necessary? • What upstream operations can affect influent water quality? • Different considerations for different uses, like potable vs irrigation
Construction/ Operation	• Who should provide construction management? • What construction areas carry the most risk? • What level of acceptance testing will be necessary? • Who will monitor water quality?
Permitting	• Who has the jurisdiction to issue the operating permit? • What are the water quality requirements? • Will the schedule be affected by the permitting? • What are the natural resources that may have an effect (land subsidence, wildlife, water sources)? • What are the environmental effects of the project and can they be mitigated?

(continued)

TABLE 2.2   Initial planning for a recycled water project (*Continued*).

Category	Challenges
Economic Effects/Costs/ Funding	• How much will the project cost? • What is the job creation potential for the project? • Are partners required and what is their participation? • Are grants available or obtainable? • Should the agency secure loans, bonds, or pay cash? • Who is responsible for ongoing operations and maintenance (O&M) and how will it be funded? • What are the triple-bottom-line benefits and life cycle costs as well as capital and O&M costs?
Public Outreach	• What is the message? • Who is responsible for sharing the message? • Supporters vs opposition
Schedule	• How long does each phase take (planning, design, construction, permitting, etc.)?

resources management. The one water approach incorporates the ideas that all sources of water and byproducts have value and should be managed in a sustainable, inclusive, integrated way. This approach recognizes that water supply, surface water, groundwater, stormwater, and wastewater are all interconnected and seeks to optimize the solutions to water resources problems for multiple purposes and multiple benefits considering the complete life cycle of water (US Water Alliance, 2016).

Under the one water approach, projects should be designed and implemented to achieve multiple benefits: economic, environmental, and social. Watershed-scale thinking and collaborative partnerships are also key points of this approach. The Water Research Foundation's *Blueprint for One Water* provides additional guidance on the phases and critical steps in developing and implementing a one water framework (WRF, 2016).

Water reuse projects can cross the typical silos of water, wastewater, stormwater, and surface water/flood protection, and require input from a diverse array of stakeholders. Often, the benefits of a water reuse project are difficult to quantify solely on an economic basis. Therefore, the one water approach provides a constructive means to evaluate potential water reuse projects in the greater context of integrated water resources.

2.2   Evaluate Opportunities

When planning for water reuse, opportunities for collaboration on water resources between water, wastewater, and stormwater utilities as well as other water stakeholders, such as those in industry and agriculture, should be explored. The needs of each stakeholder may be different and should be evaluated to determine how they may potentially benefit from the project. Planning efforts should be integrated among the various stakeholder owners or utilities to promote a common vision. It is also important that established water resources projects and associated stakeholders participate in the planning dialogue to promote water valuation that is consistent with regional and watershed sustainability.

All available sources of water should be identified, including recycled water, surface water, groundwater, seawater, and stormwater. All water users should also be identified to match users with the available sources. Embedded resources in water can be considered. For example, nutrients can be beneficial for agricultural uses and treatment processes can yield biogas for energy production and biosolids for beneficial uses. Opportunities may also exist for water trade between communities or regional users. The highest and most beneficial role of water resources should be considered under the one water approach so that quality, quantity, and location of the water sources are matched with the appropriate quality, quantity, and location of water demands.

For many applications, recycled water can be an obvious solution for areas affected by water scarcity, like the Southwest. However, in areas with seemingly large water resources such as the Southeast, mid-Atlantic, and others, recycled water adds a powerful tool to the toolbox of alternatives to alleviate the issues derived from excessive withdrawal from groundwater aquifers, which are subject to increasing limits from the local regulators, and may hinder further development in those areas.

2.3 Identify Challenges

The one water approach can produce numerous benefits for communities. However, planning for water reuse as part of one water also comes with some challenges that should be considered, including the following:

- In most regions, water, stormwater and wastewater planning is not integrated;
- Planning is most beneficial when a watershed approach is taken, but geopolitical boundaries are often not aligned with watershed boundaries;
- Regulatory structures are often in separate silos for drinking water, wastewater, stormwater, and water resources management and may have competing objectives;
- Existing economic and financial systems may create a disincentive for sharing of resources across jurisdictions; and

- Public opinions may not reflect the one water strategy, especially in areas where water scarcity is not an issue.

Keys to addressing challenges include strong leadership and vision from senior-level personnel; improving coordination between departments, organizations, and/or agencies' transparent processes and sharing of data; and building partnerships (WE&RF, 2015).

3.0 COLLABORATIVE PARTNERSHIPS AND FUNDING FOR "ONE WATER"

3.1 Exploring Partnerships and Planning Together

The one water approach will, in many cases, lead to the collaboration of different entities as partners in the planning, implementation, and O&M of water reuse projects. These partnerships are based on the grounds of mutual benefits and mutual interest. All potential partnerships should be explored and analyzed during integrated watershed planning. The various potential partners include the following:

- Utilities (private or public, all sectors);
- Agriculture;
- Power plants;
- Industry, manufacturing, and other entities with wash water or process water uses;
- Parks, golf courses, public facilities, homeowners associations, and other facilities with private irrigation;
- Environmental stakeholders; and
- Regulatory agencies.

Contracts or memorandums of understanding should be considered at the planning stage for facilitating the ease of collaborative planning and exchange of data between entities. Eventually, the partnership of different entities in the execution of a project will often involve formal contracts and, at times, the formation of legal entities.

In addition to partners that either provide, convey, and/or receive water reuse resources, other stakeholders that have an interest in the watershed should be viewed as partners as well, including nongovernmental organizations that have a particular interest in the environment. Engaging these stakeholders in the planning phase will facilitate a comprehensive understanding of the effects of the project, working ultimately toward mutual benefit and acceptance. Regulatory agencies should also be worked with in a collaborative manner and engaged during planning, especially when regulations for a particular type of reuse have not yet been established. Partnerships between utilities and customer partnerships are two particularly relevant types of partnerships that are highlighted and discussed further in the following section.

3.2   Inter-Utility Partnerships

Inter-utility partnerships are some of the most common in water reuse projects because of the existence of many utilities in the same watershed and the high potential to mutually benefit from collaboration. Inter-utility partnerships could be formed between different service sectors (i.e., water, wastewater, and/or stormwater) or multiple utilities of the same region, some of which may be of the same service sector. Operational and management efficiencies have been observed for inter-utility partnerships, in general, even outside water reuse schemes, with various lessons learned documented (U.S. EPA, 2009). Partnerships among multiple utilities in the same region may produce greater financial, social, and environmental benefits from a regional scale project and infrastructure.

An example of a common inter-utility partnership is one between a water utility and a wastewater utility. Some of the potential benefits of this partnership are as follows:

- The wastewater utility avoids the cost of discharge and avoids/offsets potential future treatment costs for nutrient/contaminant treatment;
- The water utility receives a potable water supply that is locally controlled and drought-resistant;
- Water utilities and customers receive an economically favorable, reliable water supply;
- Water utilities may be able to avoid the cost of developing new water sources to address increased water demand by substituting reclaimed water for new water where appropriate; and
- The environment may benefit from reduced discharges of pollutants in wastewater.

Different types of agreements may be made between utilities in a collaborative partnership, whether formal or informal. In the context of nonpotable recycled water systems, utilities have cited that a joint powers authority is an efficient way of establishing responsibility and authority of different partners regarding financing, constructing, operating, or maintaining a recycled water system, and that it also provides long-term institutional stability (Rosenblum, 2012). In partnerships where the delineation of responsibility is clearer and easier to define, a contract or memorandum of understanding may be sufficient. These options and others should be considered and evaluated when exploring partnerships.

3.3   Customers as Partners

Viewing customers as partners is beneficial for planning, developing, and maintaining reuse projects that provide the best value to all parties involved as well as the community and environment that is affected. Any entity that purchases water or recovered resources may be a customer, including both public and private entities. Establishing defined customers as partners will facilitate the treatment and delivery of reuse water that is fit for purpose and, therefore, help avoid costly levels of treatment when not necessary.

Agriculture, industry, and power plants are a few examples of customers that can all benefit greatly from reuse water for its potential lower cost compared to other sources as well as its reliability of supply and tailored quality. Customers and potential customers should be engaged both initially and on an ongoing basis to ensure they understand opportunities for diverse, sustainable sources of water. When the general public is a customer, they should be viewed as a partner as well. Their involvement and acceptance of the project is critical to its success, as will be discussed further in this text, particularly Chapter 6.

3.4   Challenges to Successful Partnerships

There are challenges to collaboratively working with partners, and the following are some important considerations. All of the following were identified and discussed in some form by utilities as part of a workshop on Interagency Partnerships for Water Reuse (Rosenblum, 2012):

- Differing objectives and drivers between partners need to be recognized as do mutual interests and benefits,
- Long-term relationships need to be established,
- A successful partnership must establish the responsibilities and authorities of each partner,
- Financial cost and revenue sharing need to be equitable and agreed upon, and
- Past poor relationships may exist and must be overcome.

3.5   Case Studies of Successful Partnerships

A few notable case study examples of successful partnerships are described in Table 2.3. The projects were selected to demonstrate a range of different types of projects and partnerships; many additional successful partnerships exist that are not listed here.

3.6   Funding and Partnerships

The topic of funding will be covered in depth in Chapter 5, but the following are a few specific considerations regarding funding as it relates to partner relationships:

- Financing needs to be coordinated and agreed upon between partners that will be sharing cost,
- How billing rates are set and adjusted and who has the responsibility of the administration of billing needs to be coordinated,
- Joint powers authority and other official legal partnerships may give more credibility when applying for funding (Rosenblum, 2012), and
- Having a public utility as one of the partners may allow access to funding that is not available to other entities.

4.0   PRODUCT DEVELOPMENT

A key step in planning a successful water reuse project includes identifying who the users of the recycled water end product will be, what are their needs in terms of water quality and quantity, and how to distribute the product to the users. Without customer understanding and acceptance of recycled water, the project's goals and purpose are not likely to be met.

4.1   Determine End Uses and Customers of Recycled Water

Depending on the goals of the project, the "customer" may be the utility itself or other utility agencies, agriculture, industry, institutions (university, government, etc.), recreational facilities (golf courses, parks, etc.), or the general public.

Projects should look to produce recycled water products that meet local market needs. Example end uses of recycled water are listed in Table 2.4.

4.1.1   Identify Existing Water Supplies and Users

The first step in planning should include identifying all existing water suppliers, users, and discharges in the region to determine where needs and

TABLE 2.3 Case studies of successful partnerships.

Water reuse project, location	Partners involved	Description of collaborative partnership and mutual benefits
West Basin recycled water, Los Angeles, California	• West Basin Municipal Water District • City of Los Angeles • Los Angeles County Department of Public Works & the Water Replenishment District (seawater intrusion barrier customers) • Industrial customers (refineries) • Irrigation customers	• West Basin has a contract with the City of Los Angeles as a customer of the city's wastewater effluent • West Basin has worked closely with various industrial, commercial, and municipal customers through integrated planning, which has allowed for treatment tailored to customer water quality needs • Industrial customers receive economically favorable supply to potable water that is reliable and non-competitive with potable water in a drought
Groundwater replenishment system, Orange County, California	• Orange County Water District (OCWD) • Orange County Sanitation District (OCSD) • California State Water Resources Control Board, Division of Drinking Water (DDW) • General public	• The OCSD avoided the costly challenge of building a second outfall for ocean discharge and secured a reliable increased water supply for seawater intrusion barrier water • Regional groundwater replenishment provides 19 member water agencies and their end users with reliable supply of potable water • Successful collaboration with regulators (DDW) has been crucial for implementing this project before state regulations existed for this type of reuse • OCWD and OCSD have collaborated in educational outreach to the general public, building trust with great success

TABLE 2.3 Case studies of successful partnerships (*Continued*).

Water reuse project, location	Partners involved	Description of collaborative partnership and mutual benefits
San Diego Pure Water Program, San Diego, California	• City of San Diego • California State Water Resources Control Board, DDW • General public and local businesses	• The City of San Diego has worked closely with DDW through the planning and demonstration phases, during which regulations have not been in place for this type of project • The City has engaged the public as customers and stakeholders in various ways, including tours of their demonstration facility • The City has intentionally engaged local engineering and design firms to contribute to the project design, sharing ownership of the project through collaboration • The City and public will gain a locally controlled drought-resistant water supply and avoid additional costs of water reclamation for ocean discharge
Tres Rios constructed wetlands project, Phoenix, Arizona	• City of Phoenix • Subregional Operating Group* • U.S. Bureau of Reclamation • U.S. Army Corps of Engineers • U.S. Environmental Protection Agency	• The City of Phoenix partnered with various entities to fund and construct the wetlands, which ultimately provide treatment to meet permitted discharge requirements, provide habitat for threatened and engaged species, and also a public recreation space • Effluent from the wetlands is used for crop irrigation • A nearby power plant is a customer partner, receiving water reclamation facility effluent not going to the wetlands for cooling water

(*continued*)

TABLE 2.3 Case studies of successful partnerships (*Continued*).

Water reuse project, location	Partners involved	Description of collaborative partnership and mutual benefits
Water Conserv II, Orlando, Florida	• City of Orlando • Orange County • Various agricultural and irrigation customers • Mid Florida Citrus Foundation (MFCF) • Florida Department of Environmental Protection	• Water reclamation utilities eliminated discharge to surface waters (regulatory compliance) • Regional water supply benefit of reduced aquifer usage for irrigation, and replenishment of aquifer with reclaimed water at infiltration basins, which are also environmental preserves for endangered species • Agricultural customers receive dependable, quality supply, verified by MFCF research, through contractual agreement with utilities

*Subregional Operating Group includes multiple municipalities: Glendale, Mesa, Scottsdale, and Tempe.

TABLE 2.4 Examples of end uses for recycled water.

Type of reuse demand	End uses of recycled water
Environment	• Streamflow augmentation • Wetlands restoration • Seawater intrusion barrier
Nonpotable distribution	• Agricultural irrigation • Municipal or residential irrigation • Building or industrial cooling towers • Boiler makeup • Process water • Industrial wash water • Industrial fire protection • Toilet flushing
Potable	• Indirect potable reuse • Direct potable reuse

opportunities for recycled water exist. Water supply and demand can be compared and matched in terms of geography, quantity, and quality.

4.1.2 Identify Water Quality Requirements

The selection of source water and treatment objectives for a water reuse project should be driven according to water quality needs. Likewise, the water quality requirements of the intended end use should be evaluated and considered when looking for potential customers. Utilities should work with potential customers, regulatory agencies, and stakeholders in the early stages of planning to determine the required water quality. For example, some factors for consideration include the following:

- Is nutrient removal necessary or are nutrients desirable for irrigation?
- What levels of constituents (metals, total dissolved solids, etc.) are needed by end use?
- What are the risks to the public or environment related to the use of recycled water of a particular water quality?
- If water quality is not regulated for the source and use, how will the project treatment standards be set (e.g., stormwater use for irrigation)?
- Is the use of recycled water feasible or desirable to meet a particular water demand given the required level of treatment, source water constraints, and environmental needs?

Chapter 4 addresses recycled water quality regulations and risk assessment in more detail.

4.1.3 Identify Water Demand Requirements

The customer demand requirements for product water are another important consideration. The seasonal, diurnal, and daily variations in source water and product water demands should be evaluated and compared to determine if the recycled water supply is adequate. Variations in demand can have a substantial effect on system planning as well as the capital and operating cost of a system. For nonpotable distribution systems, the most reliable customers are those that use water at a relatively consistent rate throughout the year, such as through boilers, chillers, or process water. These uses provide a baseline demand to maintain water flow through the distribution system and a consistent revenue. Uses with high seasonal peaks, such as irrigation, tend to drive up the sizing and cost of distribution infrastructure, but do not necessarily provide a consistent revenue. Uses with large variation in demands, such as landscape and golf course irrigation, also affect decisions on reuse storage, alternative water disposal during low demands, and potential supplemental water sources to handle peak demands.

Where the demand for recycled water is greater than the supply capacity, the allocation of product water should be prioritized among various users based on water supply needs, business considerations, and project drivers. For example, it may provide more overall value to supply recycled water to high-demand customers in an area where the potable infrastructure is nearing its capacity. Alternatively, the availability of recycled water in targeted locations could serve as an incentive for new development. If reduction of the nutrient loading to receiving waters is of key concern, recycled water supplied for irrigation or similar uses that are not returned to the wastewater system may be preferable, even though those uses may be seasonal.

4.1.4 Consider Recycled Water Distribution

The distribution and delivery of product water is often a key factor in the economics of a water reuse project. Identifying anchor customers with large water needs, particularly those close to the water resource recovery facilities (WRRFs), can make the difference between a project that is feasible and one that is too costly to undertake.

Installation of parallel, nonpotable ("purple pipe") recycled water distribution systems is often difficult and costly, particularly in areas that are heavily developed. However, jurisdictions can establish the requirement for new development to install dual distribution systems for nonpotable use to help build the reuse distribution system. Decentralized treatment infrastructure, such as storage and satellite or scalping facilities that treat wastewater from trunk sewers, can be considered to optimize recovery in areas that have a significant demand, but are remote to the central WRRFs.

4.2   Fit-for-Purpose

The concept of "fit-for-purpose" means matching water of a specific quality to a use appropriate for that quality. For example, a water with quality suitable for irrigation might not be suitable for industrial use as boiler feedwater. Because water can be treated to varying qualities depending on the need, WRRFs should be aware of the end use of the product water they treat. This focus on treating to the appropriate use (fit-for-purpose) ensures both sufficient treatment for public health, environmental, or product needs while also minimizing the cost of overtreating water to a quality level much different than is actually required by the end use.

There are various treatment technologies that can produce recycled water that is fit for various water quality needs, up to and including potable uses. However, not all uses require advanced levels of treatment. The selection of technical solutions for water treatment should be tailored to the end uses of recycled water to provide economic efficiency and environmental sustainability.

Dual treatment or flexible treatment train options may be used to produce varying qualities of water for the intended purposes. In many cases, advanced treatment systems, if necessary, can be added at the point of use by a specific user.

As an example, to meet the unique needs of its commercial and industrial customers, the West Basin Municipal Water District in Los Angeles County, California, produces "designer water" for industrial and irrigation use. The types of designer water produced include the following:

- Irrigation water—filtered and disinfected for industrial and irrigation use;
- Cooling tower water—tertiary treated water with ammonia and potentially phosphorous removal;
- Seawater barrier and groundwater replenishment water—secondary water, with either lime clarification or microfiltration and reverse osmosis;
- Low-pressure boiler feed water—microfiltration and reverse osmosis membranes; and
- High-pressure boiler feed water—ultrapure water treated by microfiltration membranes and passed through reverse osmosis membranes twice.

The "designer water" approach provides West Basin with operational flexibility to meet multiple needs, including those beyond the potential of common irrigation applications. West Basin's reuse program also helps ensure the economic production in the area by supplying a critical resource to several local refineries and a power generation company.

4.3   Marketing and Branding

Recycled water is a valuable product and should be portrayed as such when marketing to potential customers. When planning a project, the benefits and advantages of water reuse should be communicated to stakeholders and customers. The value of different "product" waters of various water qualities should be determined with respect to water demand and the levels of treatment, monitoring, and distribution costs associated with producing the water.

Development of a marketing and branding strategy for recycled water, as well as a value proposition to summarize why a customer should use recycled water, can be helpful in the early stages of a project. For potable reuse projects, experience has shown that public acceptance is one of the primary challenges. For these types of projects, early communication and marketing strategies are particularly important.

Chapter 6 discusses communications and outreach in more detail. In addition, references developed through research such as *Model Communications Plans for Increasing Awareness and Fostering Acceptance of Direct Potable Reuse* (WRRF, 2015) and *Marketing Nonpotable Recycled Water: A Guidebook for Successful Public Outreach & Customer Marketing* (WRRF, 2006) can provide helpful starting points.

5.0 FINALIZING THE PURPOSE AND NEEDS STATEMENT

Recycled water facilities can be implemented as a solution to the water supply needs nationwide. Wastewater is now considered a resource that should be reused and not wasted with the one water philosophy. Working together with all project stakeholders, the final purpose and needs statement can be developed to continue moving the project forward.

6.0 REFERENCES

Rosenblum, E. (2012) *Interagency Partnerships for Water Reuse: Workshop Proceedings*; WRRF 21-06; WateReuse Research Foundation: Washington, D.C.

U.S. Environmental Protection Agency (2009) *Gaining Operational and Managerial Efficiencies Through Water System Partnerships: Case Studies*; U.S. Environmental Protection Agency: Washington, D.C.

US Water Alliance (2016) *One Water Roadmap: The Sustainable Management of Life's Most Essential Resource*; US Water Alliance: Washington, D.C.

Water Environment & Reuse Foundation (2015) Institutional Issues for Integrated "One Water" Management; WE&RF SIWM2T12; Water Environment & Reuse Foundation: Alexandria, Virginia.

WateReuse Research Foundation (2015) *Model Communications Plans for Increasing Awareness and Fostering Acceptance of Direct Potable Reuse*; WRRF 13-02; WateReuse Research Foundation: Washington, D.C.

WateReuse Research Foundation (2006) *Marketing Nonpotable Recycled Water: A Guidebook for Successful Public Outreach & Customer Marketing*; WRRF 03-05; WateReuse Research Foundation: Washington, D.C.

3

Strategic Planning and Concept Development

*Allegra da Silva, Ph.D., P.E.; Joseph Griffey, P.E.; Kim Jusek, Carlos Reyes,
Emery Myers, Mia Smith, Emily Stahl, Neil Stewart, and Tom Watson*

1.0   VISION, GOALS, AND LANDSCAPE

Holistic planning and decision-making frameworks, such as triple-bottom-line, "one water", or life cycle analysis, use multiple measures of success and interesting methodologies to plan in the face of uncertainty, such as climate change, population projections, or political climate. This chapter lays out the principles involved in planning a water reuse or other alternative water source project. These principles allow water reuse planning to be transparent and inclusive, adaptive to changing conditions, and reproducible so that plans can be updated in the future. The components in planning for water reuse are summarized in Figure 3.1 and described in subsequent sections of the chapter.

For readers who want additional information about strategic planning, readers should refer to *Framework for Evaluating Alternative Water Supplies: Balancing Cost with Reliability, Resilience and Sustainability* (Water Research Foundation, estimated publication 2017), develops a decision-support framework for water utilities for integrated water supply planning.

1.1   Vision

As described in Chapter 2, water reuse planning fits in a context of integrated water resource planning, or a vision for a one water approach (see "Prep 1" in Figure 3.1). Planning must take a long-term horizon (5 to 50 years) and be grounded in commitment from leadership and a visible agency champion. It is essential that there be a cultural mindset of resource recovery and beneficial use, rather than simply "meeting the permit" or "being the least expensive option".

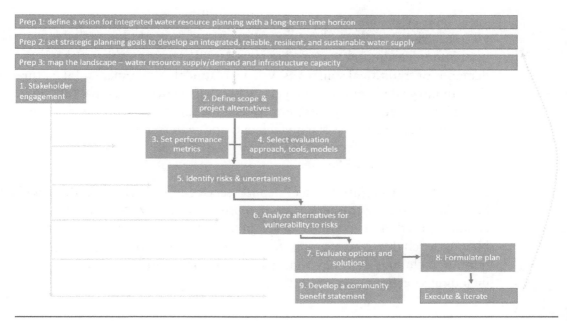

Prep 1: define a vision for integrated water resource planning with a long-term time horizon

Prep 2: set strategic planning goals to develop an integrated, reliable, resilient, and sustainable water supply

Prep 3: map the landscape – water resource supply/demand and infrastructure capacity

1. Stakeholder engagement

2. Define scope & project alternatives

3. Set performance metrics

4. Select evaluation approach, tools, models

5. Identify risks & uncertainties

6. Analyze alternatives for vulnerability to risks

7. Evaluate options and solutions

8. Formulate plan

9. Develop a community benefit statement

Execute & iterate

FIGURE 3.1   Conceptual elements of project planning.

1.2   Strategic Goals

Along with having a vision for the overall planning effort, it is critical to identify the strategic goals (see "Prep 2" in Figure 3.1). While each municipality or business that is planning a water reuse project may have specific internal goals and policies for its operation and services, very often water reuse planning goals can be distilled to the principles of developing an integrated, reliable, resilient, and sustainable water supply.

1.2.1   Integrated Water Resource Management

Achieving integrated water resource management requires looking at multiple factors affecting yield, water quality, infrastructure, and cost at once. This may include examining the use of traditional supplies such as surface water and groundwater, conservation effects, and alternative supplies such as reclaimed water, industrial wastewater, brackish water, seawater, rainwater (i.e., precipitation collected from roofs and stored in cisterns and rain barrels), stormwater (i.e., precipitation collected by storm drain systems without any engineered treatment), and gray water collected from bathing, laundry, and, in some cases, kitchen use (i.e., the non-toilet-derived portion of domestic wastewater). An integrated approach can result in lower overall capital and operation and maintenance costs while maximizing a range of social and environmental benefits. For instance, an increase in nonpotable reuse reduces demand from traditional sources while often requiring a separate distribution system (Figure 3.2). Likewise, potable reuse affects demands of traditional sources as well as treatment requirements. Taken further, an integrated approach may draw in other related municipal functions including emergency response, stormwater management, land use planning, parks and recreation, asset management, and watershed management.

1.2.2 Reliable Water Supply

Reliability is the ability to meet defined service level goals (e.g., there will be sufficient supply to service water reuse customers 95% of the time) to achieve growth and economic development of the community or industry. Having a reliable water supply takes into consideration risks and uncertainties, including shortages that may occur under future conditions, while including strategies to manage them.

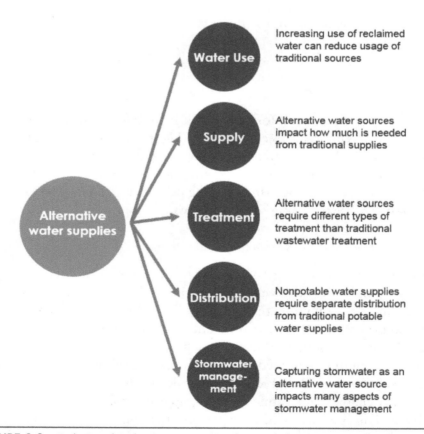

FIGURE 3.2 Relationship between increased use of alternative water supplies and traditional municipal planning components.

1.2.3 Resilient Water Supply

Resiliency refers to the ability of a water supply to recover quickly from a vulnerability or stressor (e.g., the ability to maintain service delivery during emergency events or the number of consecutive months that water quality is below a required threshold).

1.2.4 Sustainable Water Supply

For the purposes of this document, a sustainable water supply is one that meets system performance goals for reliability and resilience over a long-range future (e.g., 50 years) while minimizing triple-bottom-line effects and maximizing triple-bottom-line benefits. To achieve this, planning must include the flexibility to adapt to future conditions and include near-term, midterm, and long-term strategies.

1.3 Mapping the Landscape

A key step in laying the foundation for planning is to map the landscape by collecting data to understand the market, identify drivers, and evaluate existing and project water resource supplies and demands as well as infrastructure capacity (see "Prep 3" in Figure 3.1). This may involve a range of stakeholders.

1.3.1 Water Reuse Drivers and Regulatory Context

As introduced in Chapter 1 and again in Chapter 2, drivers for evaluating water supplies may include projections for economic development and urbanization, increasing water scarcity or changing climate conditions, evolving regulatory requirements, prevention of overdrafting of groundwater basins, intrusion of poor quality water into groundwater or land subsidence, elimination of effluent discharge, restoration of a waterbody, or a drive to diversify water resources and reduce dependency on water resources outside of local control. Regulatory guidelines and gaps should be assessed. For more discussion of regulatory considerations, refer to Chapter 4.

1.3.2 Evaluating Supplies and Demands

In this stage of preparing for planning, existing water resources and infrastructure are evaluated and projections are made to evaluate the ability of the system to meet future demands based on the drivers identified. Taking an integrated approach ensures that diverse stakeholders and existing or potential water customers are consulted, such as the agricultural community, environmental groups, and industries. The end uses, water quality requirements, and variability (seasonal and daily) of customers should be investigated. Anchor customers are identified along with existing and potential distribution systems for nonpotable reclaimed water. Finally, competing alternative sources of supply are identified.

2.0 SELECTING THE RIGHT PROJECT FOR IMPLEMENTATION

The following sections outline the steps to plan, evaluate, and select the right project for implementation. These subsections refer to the steps illustrated in Figure 3.1 earlier in this chapter.

2.1 Stakeholder Engagement

Stakeholder engagement contributes to establishing legitimacy by allowing project planners to acquire input necessary for durable, well-informed decisions and policies that reflect the unique social and environmental values of the community. Projects are best served when the process of stakeholder engagement is initiated early in project planning and integrated with overall goals rather than a stand-alone activity, as illustrated in Figure 3.1. Chapter 6 describes outreach approaches that foster transparent planning processes in greater detail. Stakeholder engagement can take many forms, from website information portals to independent advisory groups for the most complex projects.

2.2 Establish Planning Scope and Alternatives

Defining a clear scope of the planning effort is a key initial step. The following questions are addressed during this scoping and mapping exercise:

- What is the "problem" the plan is trying to solve?
- What options and alternatives should be considered in the plan?
- What should be integrated to the planning process?
- What is the right planning horizon?

2.2.1 Framing the "Problem"

Initially, it is important to define the challenge that is being addressed through the planning exercise. This is the key conclusion from the mapping the landscape preparation step, described in Section 1.3 of this chapter.

2.2.2 *Identify Potential Options to Include for Evaluation*

In this step, planners identify markets/opportunities for reclaimed water and other options that address the identified challenge, as well as applicable water quality and treatment requirements. Potential components can include the following:

- Water supply demand management (conservation);
- Stormwater and flood management and green infrastructure;
- Rainwater harvesting;
- Wastewater management, water reuse, and resource recovery;
- Water and energy efficiency;
- Environment and waterways protection; and
- Desalination.

Some of the project alternatives that can be explored are depicted in Figure 3.3; these can include both centralized and decentralized options, such as scalping plants and on-site reuse. Various options within the reuse spectrum are introduced in Chapter 1, and concepts of one water approaches are discussed in Chapter 2. Reuse and one water components add supply to a municipality's system and change infrastructure requirements and locations. They also add complexity to the decision-making process because of feedback to the supply and potable and nonpotable systems.

2.2.3 *Defining the Level of Integration that Is Needed*

Depending on the challenge being addressed in the planning exercise, different scales of integration are appropriate. For the broadest effect (and largest challenges), a watershed approach may be necessary to involve all potential

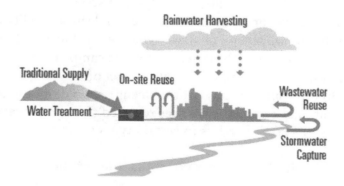

FIGURE 3.3   Illustrative types of alternative water sources that can be explored.

partners and neighboring agencies that affect watershed planning, including other water, wastewater, and stormwater utilities, as well as stakeholders from economic sectors such as agriculture, industry, developers, and power. On the other hand, for more confined challenges, integration among a handful of partners or even solely within a single agency may be sufficient. This may include stakeholders with various functional roles such as planning, regulatory compliance, engineering, operations, finance, and legal.

2.2.4   Defining the Planning Horizon and Key Decision Thresholds

Plans that require specific actionable solutions are appropriate for short-term horizons, such as 0 to 5 years, and may be incorporated to a capital improvement plan (CIP) with an associated accurate cost estimate. On the other hand, plans that need some level of specificity, but are beyond a CIP window, are midterm. Often, plans that fall in this timeframe may have "triggers" that would tip the scale toward action. For instance, if a state regulation changes, allowing a new type of reuse, a utility could begin to plan projects that leverage that new category of reuse. Long-term horizons (15 to 50-plus years) are appropriate for buildout futures and for general solutions and approaches. Long-term planning can help keep options open for the future, which may become relevant under new conditions. For example, in some communities, potable reuse is not required in the short-term or midterm timeframe, but may be necessary long-term. Investing in nonpotable reuse and associated conveyance systems may mean that potable reuse is precluded in the future because of the sunk investment. In these cases, considering the potential need for potable reuse in the long-term can affect decisions made in shorter time horizons so that the option is kept open for ultimate planning flexibility.

2.3   Set Performance Metrics

Clearly defining performance metrics or level of service goals is an important factor to account for reliability, resiliency, and sustainability of project components. Examples of level of service goals are included in Table 3.1. By establishing these goals clearly, different options can be rationally weighed against each other. For example, evaluating level of service goals for water reuse can help determine the level of treatment required or desired for new projects, operational or process changes required to provide water quality, and regulatory or permit barriers to certain types of reuse.

2.4   Select Evaluation Approach, Tools, and Models

An evaluation approach is the method to organize, conduct, and analyze the system performance under uncertainty to compare alternatives, make

TABLE 3.1   Examples of level of service goals.

Category	Example level of service goals	Example metrics
Supply	Use X% water supply sources that are naturally replenished vs Y% water supply sources that are not naturally replenished	Annual use of water supply sources that are naturally replenished (rainfall, snowmelt, runoff) vs water supply sources that are not naturally replenished (groundwater)
Treatment	Redundancy requirements	Capacity reduction when process unavailable
	Water quality targets	Number of days not in compliance
	Treatment capacity and reliability	Percentage of time capacity is insufficient
	Shutdown occurrence	Percentage of time plant is operational
	Discharge requirements (including waste streams)	Effluent (discharge) or residuals requirements
Storage	Meeting variations in demand, short-term and seasonal; resilience in recovering from a water shortage	Volume of water in storage; frequency of watering restrictions; number of customer complaints
Distribution	Hydraulic capacity; ability of system to maintain pressure while meeting hydraulic demand through demand variations	Minimum recorded pressure under varying flow conditions
Water conservation	Per capita conservation targets	Year-to-year per capita use changes; changes in peak day demand
Energy conservation	Volume based treated energy conservation targets	Energy usage
Reuse	Per capita reuse targets Water quality targets to meet needs of customers and environment	Year to year per capita use; changes in peak day demand; meet demands X% of the time (seasonal, diurnal, daily); water quality targets
Stormwater use	Stormwater use volumetric targets	Capture rate of stormwater, volume, or mass reduced from stormwater discharge

decisions, and recommended a plan. Evaluation helps managers plan, verify, and communicate what they aim to do; decide how to allocate resources; learn how best to modify or redesign programs; and estimate the resulting program outputs, outcomes, and effects. Evaluation also provides information for accountability: *Did we do what we said we would do?*

The formal process, model, and tools provide decision-makers with a consistent and tested format to structure the steps of analysis, evaluate what can be high-risk decisions with long-term effects, and can justify the selected alternative. An evaluation can use quantitative or qualitative data, and often includes both. Both types of data provide important information for evaluation and can improve stakeholder understanding and community engagement. One aspect of an evaluation approach, which assesses effects because of uncertainties and risks, is described in Sections 2.5 and 2.6 of this chapter. The evaluation approach as it is presented in this report compares alternatives first in this section, then evaluates uncertainties and risks. However, that assessment can be done concurrently or before the comparison of alternatives.

2.4.1   Overview of Evaluation Methods

The aim of this section is to provide a starting point to become aware of and to identify commonly used evaluation methods. The summary is not an exhaustive list, but serves to introduce the wide availability of evaluation methods, from those well suited for small projects with relatively easy decisions and planning efforts with few stakeholders, well-defined criteria, and certainty of information, to those designed for decisions that are more complex, involving more wide-reaching stakeholder groups with conflicting priorities and uncertain inputs and outputs.

Table 3.2 includes a description of various evaluation methods, what the method might best be suited for, whether proprietary computer software is required, potential level of training required, the range of complexity the method is suited for, level of effort or range of time commitment to use the method, as well as references for the method, including companies that may offer professional services or access to templates and examples. Complexity in decision-making not only refers to information uncertainty, the complexity of a project, or future effects, but also to the choice of assessment methodology. Section 2.4.2 highlights the decision matrix analysis, which is a less complicated and potentially less time-intensive method. Section 2.4.3 includes details of the probabilistic method, which is suited for more complex alternatives with more uncertainties, and often requires a higher level of training or a professional with experience.

TABLE 3.2 Comparison of evaluation methods (methods marked with an asterisk are described in greater detail in the following sections).

Approach	Description	Best used for / target characteristic	Proprietary computer software (yes, no)	Training required (high, medium, low)	Complexity of alternatives (high, medium, low)	Level of effort (high, medium, low)	References
Robust decisions, that is, robust decision-making or many-objective robust decision-making (MORDM)	Iterative decision analytic framework that aims to help identify potential robust strategies, characterize the vulnerabilities of such strategies, and evaluate the tradeoffs among them.	Deep uncertainty	No	Medium	High	High	https://www.google.com/webhp? sourceid=chrome-instant&ion= 1&espv=2&ie=UTF-8#q=robust+ decisions Open source MORDM: http://www .sciencedirect.com/science/article/pii/ S1364815215300190
Decision matrix, that is, decision matrix analysis (DMA) *	Decision matrix helps you to decide between several options, where you need to take many different factors into account. Simplest form of MCDA.	Well suited for approximate or subjective data	No	Low	Medium	Low	https://www.mindtools.com/pages/ article/newTED_03.htm
Multiple criteria analysis, that is, multiple criteria decision analysis (MCDA), or multiple criteria decision aid or multiple criteria decision management	Sophisticated MCDA can involve highly complex modelling of different potential scenarios, using advanced mathematics.	Can involve highly complex modeling of multiple alternatives	Software available, not required	Medium	High	High	Criterium DecisionPlus, developed by InfoHarvest Inc. (http://www .infoharvest.com/ihroot/index.asp) DecideIT, developed by Expert Choice (http://expertchoice.com/)

(continued)

TABLE 3.2 Comparison of evaluation methods (methods marked with an asterisk are described in greater detail in the following sections) (*Continued*).

Approach	Description	Best used for / target characteristic	Proprietary computer software (yes, no)	Training required (high, medium, low)	Complexity of alternatives (high, medium, low)	Level of effort (high, medium, low)	References
Probabilistic model (Monte Carlo simulation; regression models; probability trees; Markov models)*	Broad class of computational algorithms that rely on repeated random sampling to obtain numerical results. Result is a quantified probability.	For simulating phenomena with significant uncertainty in inputs. Most useful when it is difficult or impossible to use other approaches.	Software available, add-ons for worksheets available	High	High	High	Frontline Solver: Risk Solver (http://www.solver.com/monte-carlo-simulation-tutorial https://www.riskamp.com) Lumina Decision Systems: Analytica (http://www.lumina.com/technology/monte-carlo-simulation-software/) Oracle: Crystal Ball (http://www.oracle.com/technetwork/middleware/crystalball/overview/index.html) Palisade: @Risk (http://www.palisade.com/risk/?gclid=CjwKEAiAoaXFBRCNhautiPvnqzoSJABzHd6hGK-CmN-W3PzBawFpMSc7M4XjY1zaqP7ciNV63l6JxoCFmDw_wcB)
Deterministic (discrete scenarios)	May assign values for discrete scenarios to see what the outcome might be in each. Can focus on a few discrete outcomes, ignoring hundreds or thousands of others.	Addresses what exists. May examine a few, discrete outcomes such as worst, best, and most likely case scenarios	Software available, not required	Medium	Medium	Medium	https://www.vertex42.com/ExcelArticles/mc/DeterministicModel.html AnyLogic (http://www.anylogic.com/)
Triple bottom line	An accounting framework focused on three bottom lines: social (people), environmental (planet), and financial (profit).	Best used to compare different alternatives	No	Low	Medium	Medium	http://sustainability.com/who-we-are/our-story/

2.4.2 Simple, Do-It-Yourself Evaluation

Decision matrix analysis (DMA) guides a decision process between several options, where many different factors may be attributable. The DMA process is well suited for approximate or subjective data. Free worksheets, templates, and examples are available online through a range of companies and institutions and the method is fairly easy to follow in that a matrix evaluation can be quickly developed with little to no specific training other than general knowledge of spreadsheets. The steps include identifying the factors or criteria to be evaluated, scoring each choice for each factor from 0 (low) to 5 (high), and assigning weights to represent the priority or the importance of each criterion. Each score is multiplied by the weight of the factor and then total scores are summed for each option. The highest scoring option is considered to be the best alternative.

Table 3.3 illustrates a simple matrix developed for a range of criteria, the weighting factors or priority for the criteria, and total values for each priority. A decision-support tool matrix such as this can be tailored to the needs of the decision-makers, and is fast and easy to operate. Worksheets are available to download for free (Mind Tools, 2017).

2.4.3 Complex Evaluation

Higher complexity projects or alternatives with a less certain array of inputs and outputs may require a more in-depth evaluation method. Probabilistic models such as the Monte Carlo simulation represent a class of algorithms that rely on repeated random sampling to obtain numerical results to approximate the full range of possible outcomes and the likelihood of each (U.S. EPA, undated; GoldSim, 2017; Lumina, 2017). The result of a Monte Carlo simulation is a quantified probability, for example: If the water resource alternative chosen is direct potable reuse, there is a 90% chance that the water supply will reliably meet demand.

TABLE 3.3 Sample decision matrix.

Criteria	Cost	Quality of water	Public perception	Reliability of source quantity	Envision score	Project funding stability	Total
Weights	4	5	1	2	3	3	
Alternative 1	4	0	0	2	500	4	1532
Alternative 2	0	15	2	4	700	3	2194
Alternative 3	8	10	1	6	600	5	1910

There are computer software models developed specifically to conduct Monte Carlo simulation, as well as free downloads to add onto existing spreadsheets. Monte Carlo simulation is easy to use for engineers who have only a limited working knowledge of probability and statistics, and training is available.

2.4.4 Triple Bottom Line

Triple bottom line is a framework to measure criteria and performance within three parts: (1) social (people), (2) environmental (planet) or ecological, and (3) financial (profit) (NRC, 2014). "Triple bottom line" was coined in 1994 by John Elkington (1997). His argument was that companies should be preparing three different bottom lines, beyond the traditional measure of corporate profit, or the bottom line of the profit and loss account. As suggested in Figure 3.4, the overlap of social, environmental, and financial sectors creates a balanced system.

The quadruple bottom line extends the conventional three bottom lines to add a fourth part to the framework. The fourth bottom line, called *purpose*, is often expressed as spirituality or culture. For example, in relation to a water reuse project or alternative under evaluation, the fourth bottom line might answer what is the purpose of this project or alternative.

Organizations and businesses have adopted the triple-bottom-line framework to evaluate their performance, accounting for the full cost involved in doing business. More information on triple-bottom-line analysis is provided in Chapter 5. The triple-bottom-line framework has migrated from an organizational evaluation to include a method of evaluation of alternatives and projects, for instance, in the Envision sustainable infrastructure rating system (ISI, 2017).

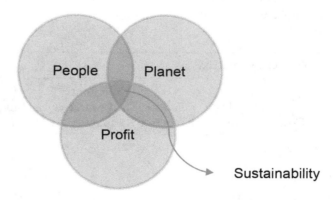

FIGURE 3.4   Conceptual illustration of triple-bottom-line components.

Envision is a system for rating infrastructure projects based on the triple bottom line for a project's overall contribution to the economic, environmental, and social aspects of sustainability as well as the fourth bottom line of purpose. Envision not only asks, "Are we doing the project right?", but also, "Are we doing the right project?", implying understanding of purpose. Envision can be used as a system to pick the right project at the outset. When used in the earliest planning phases, it can help identify options that result in significantly better outcomes.

Envision is a holistic sustainability rating system for all types and sizes of civil infrastructure, including the roads, bridges, pipelines, railways, airports, dams, levees, landfills, water treatment systems, and other components that make up civil works. It is a system to estimate the sustainability of a project or alternative, ranging from "bronze" (low sustainability) to "platinum" (high sustainability). The Envision rating system is an example of multicriteria decision analysis and scoring criteria to help identify and measure variables that reflect social, environmental, and financial effect categories and compare alternatives numerically. Envision can be used as a combination analysis in conjunction with other tools or analysis methods.

The rating system uses a score based on a point system for a range of credits defined in five categories: (1) quality of life, (2) leadership, (3) resource allocation, (4) natural world, and (5) climate and risk. Each of the credits and their associated levels of achievement are assigned points weighted in accordance with their estimated contribution to sustainability. *The Envision Guidance Manual* (ISI, 2015) summarizes each credit description and provides guidance to determine the level of achievement for a given project. Credits can be omitted if it can be shown that they are not applicable to the project. Figure 3.5 illustrates the five categories and the credit list developed for the rating system.

Figure 3.6 illustrates the levels of achievement for the scoring system, ranging from "improved", a low level of achievement, to "restorative", a high-level achievement and higher score.

After evaluation of alternatives or a project evaluation during a life cycle phase, the application of the rating system may be reviewed by a third-party verifier and authenticators. These are trained professionals with the Envision rating system who confirm levels of achievement and provide quality control of the system. The verification and authentication steps are required if the project under question is to be assessed for an Envision award. Otherwise, the rating system may be used as an internal tool for the team conducting the alternatives analysis or to optimize design, construction, or operations activities.

QUALITY OF LIFE
13 Credits

1 PURPOSE
QL1.1 Improve Community Quality of Life
QL1.2 Stimulate Sustainable Growth & Development
QL1.3 Develop Local Skills and Capabilities

2 WELLBEING
QL2.1 Enhance Public Health and Safety
QL2.2 Minimize Noise and Vibration
QL2.3 Minimize Light Pollution
QL2.4 Improve Community Mobility and Access
QL2.5 Encourage Alternative Modes of Transportation
QL2.6 Improve Site Accessibility, Safety & Wayfinding

3 COMMUNITY
QL3.1 Preserve Historic and Cultural Resources
QL3.2 Preserve Views and Local Character
QL3.3 Enhance Public Space

QL0.0 Innovate or Exceed Credit Requirements

LEADERSHIP
10 Credits

1 COLLABORATION
LD1.1 Provide Effective Leadership & Commitment
LD1.2 Establish a Sustainability Management System
LD1.3 Foster Collaboration and Teamwork
LD1.4 Provide for Stakeholder Involvement

2 MANAGEMENT
LD2.1 Pursue By-Product Synergy Opportunities
LD2.2 Improve Infrastructure Integration

3 PLANNING
LD3.1 Plan for Long-Term Monitoring & Maintenance
LD3.2 Address Conflicting Regulations and Policies
LD3.3 Extend Useful Life

LD0.0 Innovate or Exceed Credit Requirements

RESOURCE ALLOCATION
14 Credits

1 MATERIALS
RA1.1 Reduce Net Embodied Energy
RA1.2 Support Sustainable Procurement Practices
RA1.3 Use Recycled Materials
RA1.4 Use Regional Materials
RA1.5 Divert Waste from Landfills
RA1.6 Reduce Excavated Materials Taken Off Site
RA1.7 Provide for Deconstruction and Recycling

2 ENERGY
RA2.1 Reduce Energy Consumption
RA2.2 Use Renewable Energy
RA2.3 Commission and Monitor Energy Systems

3 WATER
RA3.1 Protect Fresh Water Availability
RA3.2 Reduce Potable Water Consumption
RA3.3 Monitor Water Systems

RA0.0 Innovate or Exceed Credit Requirements

NATURAL WORLD
15 Credits

1 SITING
NW1.1 Preserve Prime Habitat
NW1.2 Protect Wetlands and Surface Water
NW1.3 Preserve Prime Farmland
NW1.4 Avoid Adverse Geology
NW1.5 Preserve Floodplain Functions
NW1.6 Avoid Unsuitable Development on Steep Slopes
NW1.7 Preserve Greenfields

2 LAND + WATER
NW2.1 Manage Stormwater
NW2.2 Reduce Pesticides and Fertilizer Impacts
NW2.3 Prevent Surface and Groundwater Contamination

3 BIODIVERSITY
NW3.1 Preserve Species Biodiversity
NW3.2 Control Invasive Species
NW3.3 Restore Disturbed Soils
NW3.4 Maintain Wetland and Surface Water Functions

NW0.0 Innovate or Exceed Credit Requirements

CLIMATE AND RISK
8 Credits

1 EMISSIONS
CR1.1 Reduce Greenhouse Gas Emissions
CR1.2 Reduce Air Pollutant Emissions

2 RESILIENCE
CR2.1 Assess Climate Threat
CR2.2 Avoid Traps and Vulnerabilities
CR2.3 Prepare for Long-Term Adaptability
CR2.4 Prepare for Short-Term Hazards
CR2.5 Manage Heat Island Effects

CR0.0 Innovate or Exceed Credit Requirements

FIGURE 3.5 Envision rating system categories and credit list (reprinted with permission from the Institute for Sustainable Infrastructure).

2.5 Identify Uncertainties and Risks

The decision frameworks described in Section 2.4 help to systematically evaluate project opportunities. All of the frameworks incorporate an evaluation of uncertainty and risk. Uncertainties include future conditions that are unknown or cannot be accurately estimated. On the other hand, risks are able to be reasonably estimated. The objective of this stage in project selection is to identify uncertainties and risks to the water supply system or project and what assumptions the organization makes regarding them, to prioritize them, and to identify which should be analyzed further. Uncertainties and risks may include factors such as the following:

LEVELS OF ACHIEVEMENT

FIGURE 3.6   Envision levels of achievement for rating criteria (reprinted with permission from the Institute for Sustainable Infrastructure).

- Supplies—How may future water supplies be affected by political decisions, water rights issues, and other socio-political effects? How may drought, climate variability, or other types of hydrologic variability influence water supply, storage, and quality?

- Demands—How will the changing population, local economy, and urban land use and growth patterns drive demands for alternative water sources and one water approaches? Which industries will become key partners in the one water puzzle?

- Institutional—Will other city and regional agencies be willing partners in implementing one water concepts?

- Social—What are the political, social, or other barriers to proposed options and how much of an effect do these have? How may social values related to water use change?

- Regulatory—How will future regulations affect allowable uses of reclaimed water and the viability of alternative water supplies and uses? What regulatory gaps exist for this project? Who will determine design standards?

- Innovation and execution—What technological or economic innovations will advance alternative water sources, including for large centralized treatment and monitoring as well as on-site reuse? Are there any risks to early technology adoption?

These factors are summarized pictorially in Figure 3.7.

2.6   Analyze Alternatives for Vulnerability to Risks

After the uncertainties and risks from Section 2.5 have been identified for different options, the vulnerability of the options to the uncertainties and risks needs to be assessed. There are both qualitative and quantitative approaches for performing this assessment.

Qualitative approaches score uncertainties and risks based on categories such as likelihood and effect. In this approach, each uncertainty and risk is assigned a score for relative likelihood and a score for relative impact. For example, scores may range from "1" (extremely rare likelihood/minimal impact) to "5" (certain likelihood/critical impact). The individual and composite scores are then used to determine the uncertainties and risks to which the system is vulnerable, as shown in Table 3.4. For example, one potential risk for many reuse options is the regulatory framework that affects their feasibility. For different reuse options, the effect of a regulation change may be different, suggesting different vulnerabilities to this risk. These qualitative

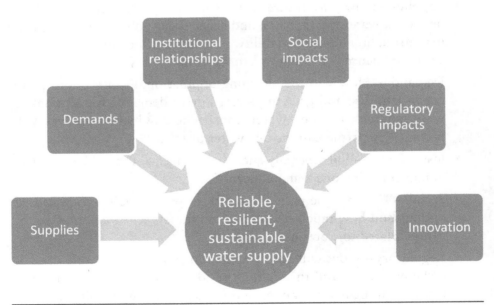

FIGURE 3.7   Key categories of uncertainties affecting water utilities.

TABLE 3.4 Conceptual ranking for scoring uncertainties and risks.

Score	Likelihood	Effect
1	Rare	Insignificant
2	Unlikely	Minor
3	Possible	Moderate
4	Likely	Major
5	Certain	Extreme

methods are based on perceptions by subject matter experts, and thus are biased based on their opinions.

Quantitative approaches map impacts from identified uncertainties and risks to the options using modeling to identify the uncertainties and risks to which the options are vulnerable. Quantitative approaches fall into two broad categories. One type modifies the system inputs and measures the magnitude of the impact from an uncertainty or risk (e.g., percentage reduction in capacity because of an outage) until the magnitude is high enough to cause performance degradation. This method is effective for uncertainties when the impact is unknown or difficult to estimate. The other approach assumes a set impact from an uncertainty or risk and applies it to the option, evaluating whether the impact is significant enough to cause performance degradation. Both of these methods can evaluate each uncertainty and risk individually, or in combination with others. Quantitative methods have the benefit of reducing the influence of bias. However, this approach is more computationally complex, requiring a computer model of the system and the ability to conduct many simulations.

2.7 Formulate a Plan

As the final step in the process, the evaluation criteria described in Section 2.4 are combined with the vulnerability assessment results from Section 2.6 to select a portfolio of options. Ideally, this portfolio contains options that have low vulnerability, score highly on the evaluation criteria, and have low cost. In almost all cases, however, there is no single perfect portfolio, but rather multiple portfolios that trade off cost, performance, and triple-bottom-line criteria. For example, the best performing portfolios may be the most expensive, whereas the least expensive portfolios may have poor performance. Therefore, portfolios that highlight these trade-offs should be identified, giving decision-makers the opportunity to weigh them and to ultimately decide which tradeoff provides the most desirable outcome.

After the portfolio of options is selected, an implementation plan needs to be developed. Because the time between a planning process and project completion can be decades, the assumptions made during the plan development may not represent actual future conditions. To account for this, an implementation plan should be adaptive to these changing conditions. Identifying the key assumptions made about future conditions highlights the trends or "signposts" that should be monitored. For each of these signposts, identify what the impact to the plan recommendations is if the assumption is inaccurate. Then, recommendations from the plan can be modified appropriately. Table 3.5 lists some common signposts from reuse projects and potential effects on plan recommendations.

TABLE 3.5 Illustrative examples of signposts to monitor that can trigger new evaluation of alternative water sources.

Example signpost	Potential assumption made for plan	Trends to monitor	Effects on plan recommendations
Patterns of water usage in service area or facility	Amount of water demand that could be met by alternative water sources	Changing patterns of water usage that open up the possibility for increased demand for alternative water sources	Changes in demand for alternative water sources may increase or decrease the size of reuse facilities required
Regulations governing alternative water sources	Regulatory framework for reuse will be in place by a certain year	Status of new regulations	Changes in timing to regulations could delay when reuse can be completed
Technologies to treat alternative water sources	Available reuse technology will be similar to what is currently available	Emerging reuse technology	Game-changing advancements in reuse that fundamentally change a constraint may alter the feasibility of reuse
Public perception	The public is currently not open to certain alternative water sources, but is accepting of existing uses of reclaimed water (e.g., irrigation)	News media coverage or polling of public opinion for or against alternative water sources, including reclaimed water	Changes in public opinion can either increase demand for alternative water sources or potentially jeopardize existing programs where they exist

The time to complete the planning process can vary from a few months to a few years, depending on the complexity of the desired plan. Typically, the more people (both within the organization and external stakeholders) involved in the process, the longer the planning process should be expected to take. The planning process can be broken up into four phases: identifying potential options, evaluating options, deciding on portfolios, and developing an execution plan. The evaluation and decision part of the process should be expected to take the longest. For example, potential options can be identified, typically over the course of a few weeks, while evaluating options and deciding on portfolios can take several months.

2.8   Develop Community Benefit Statement

As a component of communication of the decision-making process result, a key initial step is the development of a community benefit statement. While this section is focused on a municipal audience, the concepts apply whether for articulating community benefits or business case benefits, and are, therefore, adaptable for the private sector.

The development of alternatives, subsequent alternatives analysis, and resulting communication efforts, including a project's community benefit statement, should be framed in such a way to facilitate project legitimacy, not merely public acceptance. Legitimacy is a generalized perception or assumption that the actions of an entity are desirable, proper, or appropriate within some socially constructed system of normal values, beliefs, and definitions (Ginzel et al., 1992; Perrow, 1970).

Statements that do not address project legitimacy create noise inside the larger communications strategy and run the risk of being received as self-serving or disingenuous. But, a benefit statement that targets one or more levels of legitimacy fosters trust between the agency and audience because it does one of three things: it communicates a tangible benefit for the audience, it responds to a larger need of the entire community, or it demonstrates the agency's ability to solve a real problem. Community engagement does not end with the community benefit statement; rather, it is merely an initial, essential step. Chapter 6 provides more information on the overall goals and stages of communication and outreach.

The three examples of community benefit statements that follow demonstrate the power in focused messaging aimed at establishing project legitimacy, and, in some instances, capitalize on existing legitimacy the agency already holds. The varied approach of the three statements highlights the fact that effective statements can be crafted in many different ways using different words and different themes. To a certain degree, this must be the case. If our goal is project legitimacy, this active approval is dependent on acceptance within the social constructs and the value and belief systems of the communities where the projects are proposed. Because these characteristics can vary by region, to be effective, these approaches must be influenced by and take on the identity of the local culture. As such, the statements constructed for your project may not look or sound exactly like the examples presented herein.

With that said, the following statements reflect consistency in several key areas outlined above. Although unique in their approach, all three operate in the realm of communicating big ideas. They avoid the appearance of being delivered from a position of intellectual superiority. None burden their audience with a litany of highly technical details or industry-specific jargon. And, in each instance, the language is straightforward, concise, and meaningful. Communicating in such a way is challenging, but the traction gained in having the community actively behind your project will far exceed the effort expended.

2.8.1   Orange County Water District, California, Community Benefit Statement

"In the mid-1990s, the Orange County Sanitation District (OCSD) faced the possibility of having to build a second ocean outfall that would have cost approximately $200 million. At the same time, the Orange County Water District (OCWD) was faced with continued problems of seawater intrusion and the need to expand its Water Factory 21 (WF 21) from 22.6 million gallons (85,600 cubic meters) per day to 35 million gallons (132,500 cubic meters) per day.

At the time, California had just experienced a severe drought. Water experts also projected droughts would occur three out of every 10 years, that there would be increases in demand due to population growth, and that the demand and cost of imported supplies would increase in the near future. Faced with these future challenges, OCWD built upon its long history of successfully treating wastewater at WF 21 for its seawater barrier and decided to implement advanced processes to purify the wastewater and send it to recharge basins, where it would ultimately become part of north and central Orange County's drinking water supply" (OCWD, undated).

The following are quick statement highlights:

- The problem to be solved is framed by three simple, but powerful ideas—drought, population growth, and cost of raw water supplies. These big ideas are easy for OCWD's audience to latch onto.
- The statement touches on the long and successful history of OCWD's existing reuse operations. This phrase communicates the pragmatic and moral legitimacy of OCWD, and suggests the cognitive legitimacy

already established by the agency. In short, what OCWD has done so well in reuse water in the past is a powerful antidote to any objections of what they will do with it in the future.

2.8.2 Guelph, Ontario, Community Benefit Statement

"These days, who doesn't want to make "green" choices and have smaller bills? When it comes to purchasing a new home and committing to environmental conservation, there are so many options and choices; so it's important to know what's reputable, high quality, and provides long-term environmental and financial benefits.

Home greywater reuse systems collect greywater from household showers and baths, purify the greywater collected using chlorine, and utilize the treated greywater to flush toilets within the home. Toilet flushing represents about 30 percent of daily water use in the home. Using reclaimed greywater from showers and baths eliminates the use of potable water for toilet flushing, resulting in significant annual water and wastewater cost savings" (City of Guelph, 2017).

The following are quick statement highlights:

- As part of Guelph's Blue Built home program, the implementation of gray water reuse systems in new home construction is paired with the larger "green" movement prevalent in many social circles, especially among millennials. For these young individuals that may be looking for a way to marry their environmental conservationism with their first home purchase, this social relevance lends moral legitimacy to point-of-use gray water reuse systems.

- The statement acknowledges the challenge in evaluating the claims of a whole host of products and systems that claim environmental friendliness. In doing so, the statements suggest a partnership between agency and audience, and together they will arrive at the best course of action.

- Although the statement communicates a financial benefit of the gray water reuse system, the prevailing theme is environmental stewardship.

2.8.3 San Diego, California, Community Benefit Statement

"San Diego relies on importing 85% of its water supply from the Colorado River and Northern California Bay Delta. The cost of this imported water has tripled in the last 15 years and continues to rise. With limited local control over its water supply, the City of San Diego is more vulnerable to droughts, climate change and natural disasters.

The Pure Water Program:

- Uses proven technology to clean recycled water to produce safe, high-quality drinking water
- Provides a reliable, sustainable, water supply
- Offers a cost-effective investment for San Diego's water needs

With San Diego's existing water system, only 8% of the wastewater leaving homes and businesses is recycled; the rest is treated and discharged into the ocean. The Pure Water Program transforms the City's water system into a complete water cycle that maximizes our use of the world's most precious resource—water" (City of San Diego, 2016).

The following are quick statement highlights:

- The statement clearly describes the problem that will be solved by implementing a reuse project and the financial impact on ratepayers if no action is taken by the agency. Although not directly stated, the cost-of-service benefit to be gained by the audience is clear and pragmatic legitimacy is established.
- The project is presented as responding to a water supply issue, not a wastewater disposal issue. This is significant because previous research suggests the general public is more supportive of efforts to maintain water supplies than they are of efforts to dispose of treated wastewater.
- The message communicates three key ideas to establish moral legitimacy—proven technologies, sustainability, and cost-effectiveness.
- Although economics is specifically noted multiple times, the statement emphasizes the importance of autonomy and water independence, a benefit that may be more important to the community than the money it will save.

2.9   Execute and Iterate

Once a plan has been developed, the final step is to execute the plan recommendations. When executing a plan, the key objective is to develop a process by which short-term investment decisions are evaluated in light of long-term planning strategies. The challenge with this objective is that short-term budgeting decisions are driven by immediate capital needs based on current conditions, whereas long-term planning strategies many times do not have a capital need for years and are based on uncertain assumptions about the future. Therefore, to best blend the two, budgeting decision criteria should be developed that are informed by the organization's long-term planning strategies and concepts of reliability, resilience, and sustainability.

The planning process should be repeated every 5 to 10 years to account for changing conditions and new technologies and to re-evaluate previous recommendations. In many states, the frequency of planning updates is mandated by policies or regulations. In this kind of planning, the organizational learning and insight that are gained are often as important as the plan itself; therefore, it is important to repeat the process frequently.

3.0   CASE STUDIES

3.1   Guelph, Ontario: Planning for the Future—A Phased Approach to Integrating Water Reuse

The City of Guelph (Figure 3.8) is located in southwestern Ontario, Canada, and currently has a population of approximately 130,000 people, which is projected to grow to a total population of approximately 191,000 by 2041 (AECOM and Golder Associates, 2014). Guelph is an example of a city that has been implementing an incremental, phased approach toward planning and conceptualizing the incorporation of wastewater reuse as part of their long-term water management strategy.

FIGURE 3.8   City of Guelph water tower (reprinted with permission from City of Guelph).

3.1.1 Drivers for Water Reuse

The City of Guelph is one of Canada's largest municipalities that relies on groundwater for its water supply. The city's wastewater discharges to a sensitive river, with limited and regulated assimilative capacity. Thus, both water conservation and high-quality wastewater treatment are priorities in light of a growing population. A strong sense of environmental activism and community preference for sustainable, local supply options have also served as drivers for the city to consider reuse. The community also noted through the recent development of the *Water Efficiency Strategy* that innovation is a high priority when looking at water efficiency and conservation measures (C3 Water and Gauley Associates, 2016). Furthermore, with recent years of drought, there is additional pressure to look to alternative sources of water for tasks such as lawn watering and irrigation. Thus, it was important for the city to consider reuse initiatives.

3.1.2 Project Mapping, Evaluation, and Selection

The pathway toward reuse began with a focus on water demand management through the implementation of aggressive water conservation and efficiency programming. Since the development of the award winning *Water Conservation and Efficiency Strategy* in 1999, Guelph has become a leader in water conservation and progressive water management in Canada, with an average single-family residential demand of 165 L/cap·d) compared to a nationwide average of 251 L/cap·d in Canada (2011) and 379 L/cap·d in the United States (2016) (C3 Water and Gauley Associates, 2016; Environment Canada, 2016; USGS, 2016).

Through council-approved master planning processes, complementary studies, and the implementation of efficiency programs, Guelph has been gathering information relevant to the preferred reuse configuration and mainstreaming the concepts of reuse. As a result, many different stakeholders have been involved in the discussion around water reuse for Guelph, including the wide range of city departments, the public, city council, academia, and local consultants. One of the larger setbacks to the incorporation of reuse is the fact that municipal-scale reuse programs are not currently commonplace in Ontario and there is limited regulatory guidance on the design and implementation of water reuse.

3.1.3 Long-Term Planning

Two of the key and complementary strategic planning documents for water management in the city are the *Water Supply Master Plan* (WSMP) and the *Water Efficiency Strategy* (WES), which are updated approximately once every 5 years (City of Guelph, 2013, 2016). Long-term water supply planning and reduction targets (25 years and beyond) are determined through a public, council-approved WSMP process to ensure a continued safe, sustainable water supply (City of Guelph, 2013). The WES details water efficiency policies and programs to meet the WSMP demand reduction targets incorporated to the WSMP demand projections (City of Guelph, 2016). Building on the previous WES, the 2014 WSMP Update included a 13% (9147 m^3/d) demand reduction over 25 years (AECOM and Golder Associates, 2014; C3 Water and Gauley Associates, 2016).

Even with planned conservation, the 2014 WSMP Update identified that new water sources were required to meet projected demands around 2025 (AECOM and Golder Associates, 2014). Options evaluated included conventional supply sources such as groundwater and surface water alongside innovative options such as wastewater reuse, both as nonpotable and potable supply configurations. Water reuse was not identified as a preferred recommendation in the WSMP for the 25-year timeframe while more affordable sources are available. However, it was recommended that reuse continue to be discussed in planning documents such as the WES, re-evaluated with assessments for wastewater treatment capacity upgrades, and reviewed for opportunities to incorporate to future developments and road works (dual plumbing systems, purple pipe, etc.) (AECOM and Golder Associates, 2014). In planning beyond 25 years, the WSMP acknowledges that the feasibility of reuse is expected to increase as more advanced wastewater treatment is needed to meet discharge requirements, local freshwater resources approach their sustainable capacity, and costs associated with reuse decrease (AECOM and Golder Associates, 2014).

The 2016 WES included decentralized reuse to contribute to the WSMP reduction targets in the form of the Blue Built Home Water Efficiency Standard and Rebate Program (Figure 3.9). This program includes subsidies and technical guidance for gray water reuse ($1,000) and rainwater harvesting ($2,000), and has been an important form of public outreach for reuse (C3 Water and Gauley Associates, 2016). The WES indicates that larger-scale, centralized or semicentralized reuse schemes at municipal facilities are anticipated to have greater potential for freshwater demand reductions. Because Guelph has relatively high-quality, tertiary-treated wastewater effluent, urban nonpotable applications for reclaimed water are seen as a probable reuse scenario.

FIGURE 3.9 City of Guelph Blue Built Home Certification Program emblem.

This is attributable, in part, to the more limited additional treatment anticipated to meet industry standard guidelines for safe reuse (CH2M Hill, 2009; Genivar, 2012). The 2016 WES allocated a research budget to conceptualize and consolidate the vision for larger-scale reuse to be carried forward as a water-use reduction for the 2022 WES update. The consideration of reuse in the council-approved WSMP and WES is significant because it signals to the community the city's intention to continue to explore this topic.

3.1.4 Complementary Efforts

With growing community interest, considerations for reuse started being incorporated to a range of long-term master planning processes in the mid-to-late 2000s. Following these more formal milestones for reuse planning, opportunities and constraints for reuse have been assessed in more detail in a range of complementary studies. For example, after the concept of a purple pipe network was introduced in the 2008 *Water and Wastewater Servicing Master Plan*, a more detailed purple pipe feasibility study was completed as part of an environmental assessment for the reconstruction of sewer and watermains along a major city roadway (AECOM and Golder Associates, 2014; Genivar, 2012). Another more recent example was a study on the effect of residential softener salt on the environment and urban water cycle, which included a question on the effect on future reuse. This study highlighted that one of the likely applications of reclaimed water by the city—irrigation water—may be less feasible because of salt loading without advanced water treatment (Stantec, 2016). Because the practice of reuse necessitates a holistic, long-range water management approach, these examples highlight that considerations for reuse are necessary to include in water management decisions well in advance of its implementation.

3.1.5 Community Benefit Statement

The messaging for reuse is part of the conversation around the approach to long-term supply and the rationale for water conservation. In concert with community feedback, the city emphasizes that the future of water efficiency will be innovative and consider new technology. Important concepts that are highlighted for reuse include (1) the fact that only a portion of water uses require potable quality while currently all water produced by the city is fit for consumption, (2) as per the Ontario Building Code, rainwater harvesting and gray water reuse systems can meet some of that nonpotable demand such as toilet flushing (Figure 3.10), and (3) larger reclaimed water systems will result in larger economic returns (C3 Water and Gauley Associates, 2016). Refer to Section 2.8.2 for more information on the community benefit statement.

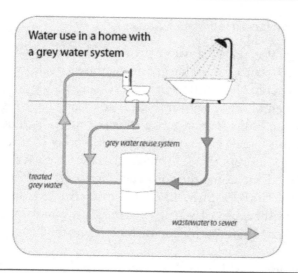

Water use in a home with a grey water system

grey water reuse system

treated grey water

wastewater to sewer

FIGURE 3.10 Ontario Building Code allows gray water reuse for toilet flushing.

3.1.6 Recap and Ongoing Challenges

The city has formally evaluated reuse as a long-term supply option in its WSMP and, while it is not presently assessed as feasible in the near-term, the public is very committed to this alternative. To further explore this option, the city has allocated budget to research and arrive at a more solidified concept for reuse in the city, which can be carried forward in the next long-term conservation strategy. Alongside this, city managers are incorporating considerations for reuse to a range of complementary studies that may provide deeper understanding of opportunities and limitations for reuse. Moving forward, answering key questions regarding the economics, environmental effects, policies, and social considerations will require coordination across water disciplines. One of the key challenges for implementing reuse in the city includes addressing and overcoming regulatory gaps. Despite uncertainties, the city has managed to set a strategic vision and move forward on conceptualizing how reuse will be implemented to meet its local water management needs.

3.2 Las Virgenes, California: Moving Toward Indirect Potable Reuse

The Las Virgenes Municipal Water District (LVMWD) and the Triunfo Sanitation Districts jointly own and operate the Tapia Water Reclamation Facility (TWRF) through a Joint Powers Authority (JPA). Located between Malibu and Calabasas in unincorporated Los Angeles County, California, the TWRF produces roughly 24.6 ML/d (6.5 mgd) of Title 22 recycled water for local residents. However, because of seasonal fluctuations in the demand, only 60% of that recycled water is beneficially reused.

3.2.1 Drivers for Increased Water Reuse

During the winter, the recycled water produced by the TWRF exceeds local demand and the balance is discharged to Malibu Creek, which eventually drains into the Pacific Ocean. Recently, increasing regulatory requirements and nongovernmental organization (NGO) advocacy have resulted in more stringent discharge limits, which make the continual discharge of recycled water into Malibu Creek excessively costly. Since 1973, the JPA has considered options to address this seasonal imbalance between recycled water supply and demand, including the construction of a new reservoir within the JPA's service area. Until recently, all efforts to realize seasonal storage have been unsuccessful because of sensitive ecological conditions, stakeholder and NGO activism, and a multitude of other factors.

3.2.2 Project Mapping, Evaluation, and Selection

To forge a path forward and find consensus among the varied stakeholders that contribute to project success, a facilitation process was initiated amongst the agency staff and key stakeholders in the area, including NGOs. This facilitation process began in 2014 with four public workshops, activities structured to solicit input from the stakeholders, and presentations from technical experts. During this first phase, six conceptual project alternatives were defined with stakeholder input. From the six alternatives, the JPA board selected two scenarios for further investigation: (1) advance-treat the TWRF effluent for indirect potable reuse (IPR) using LVMWD's water supply reservoir and (2) seasonal recycled water storage at the nearby Los Angeles Department of Water and Power's existing Encino Reservoir. Through another process of stakeholder engagement, four public workshops, and technical analysis, these conceptual scenarios were further refined into two project alternatives with strong stakeholder support. Both alternatives more efficiently address the JPA's seasonal water management issues to make full use of their recycled water resource and show promise thanks to the JPA's proactive stakeholder engagement process.

Between the two final alternatives, the JPA is actively pursuing the advanced treatment of TWRF effluent for IPR. Under this alternative, a new advanced water treatment (AWT) facility is planned that will produce up to 22.7 ML/d (6 mgd) of potable water. This water will be sent to the Las Virgenes Reservoir (Figure 3.11) for surface water augmentation and eventually to the potable distribution system. This option had several advantages. First, it created a higher end use for the district's recycled water by producing potable water. Secondly, it allowed the JPA to have autonomy over its system. Cost and schedule considerations also favored the AWT option, although the district is reserving the option to pursue seasonal storage in the Encino Reservoir. To further develop the AWT alternative, the JPA is in

FIGURE 3.11   Las Virgenes Reservoir (reprinted with permission from Las Virgenes MWD).

the process of undergoing reservoir modeling efforts, site selection for the AWT facility, and predesign for a demonstration AWT plant.

3.3   Santa Monica, California: Sustainable Water Infrastructure Project

The City of Santa Monica distributes potable water to approximately 18,000 metered customers within the city's boundaries. Currently, the water supply consists of a mixture of local groundwater and environmentally costly imported water from northern California and the Colorado River Project.

3.3.1 Drivers for Increased Water Reuse

The city has set a goal to achieve sustainable water independence from imported water by 2020. To meet this challenge, the city must find an additional forecasted supply of up to 17 ML/d (4.5 mgd or 5000 ac-ft/yr). To close the forecasted demand gap while simultaneously improving drought resiliency and the long-term yields of its aquifers, the city will continue to promote and enhance its ongoing conservation programs, identify untapped groundwater resources, and assess creative ways to treat and reuse the water that is now available.

3.3.2 Long-Term Integrated Water Resources Planning

The city has planned a long-term water solution that integrates the reuse of all available nonconventional water resources including recycled municipal wastewater, stormwater runoff, and brackish groundwater. The city's Sustainable Water Infrastructure Project (SWIP) will harvest these three valuable resources to produce approximately 5.7 ML/d (1.5 mgd or 1680 ac-ft/yr) of advanced treated water for immediate nonpotable reuse and, when properly permitted, IPR via aquifer recharge (Figure 3.12). The SWIP takes a forward-thinking approach to help secure the city's water future by linking together the three distributed water reuse elements via smart technology into a single, cohesive, and comprehensive project approach. Multiple benefits derived from the SWIP include improved beach water quality from reduced stormwater discharges to the ocean, compliance with local water board non-point-source pollution control requirements, and comportment with the state water board policy of increasing local recycled water reuse. The SWIP also addresses the key water reuse objectives in the California Water Action Plan.

The SWIP includes the construction of two large, below-grade stormwater harvesting tanks with a combined capacity of 17 ML (4.5 mil. gal) and the SWIP Recycled Water Treatment Facility (SRWTF). The 3.8-ML/d (1.0-mgd) SRWTF will treat a blend of municipal wastewater and stormwater for eventual IPR. In addition to these new assets, the SWIP will leverage the city's existing infrastructure through planned upgrades of its Santa Monica Urban Recycled Runoff Facility (SMURRF). The SMURRF currently treats urban runoff and some stormwater flows for nonpotable recycled water uses. Under the SWIP, the SMURRF will be upgraded with an off-the-shelf reverse osmosis system to treat brackish groundwater and, most importantly, produce an effluent quality that matches that produced by the new SRWTF. This will allow for blending of the two effluents, enabling the SMURRF to discharge into the same distribution and groundwater recharge

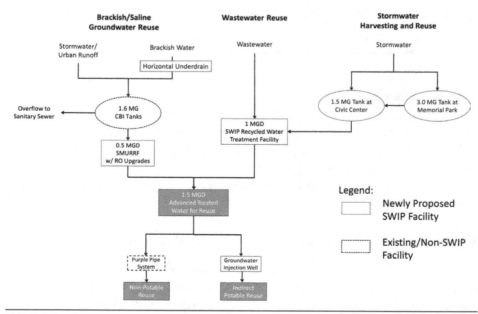

FIGURE 3.12   Overview of the City of Santa Monica SWIP.

infrastructure that will be constructed for the SRWTF. By incorporating its existing infrastructure, the city was able maximize resources and create a comprehensive water reuse solution.

3.3.3 Site Selection

In addition to the overall treatment strategy, planning efforts for the SWIP included an extensive alternatives evaluation in which the city evaluated the feasibility of six potential sites for the SRWTF. The sites were evaluated for constructability, compatibility with the surrounding area, cost, and schedule. Because of concerns related to community acceptance and the potential loss of valuable surface space and parking in the city, all of the SWIP infrastructure, with the exception of the upgrade of the existing SMURRF, will be constructed below grade. This innovative approach allows for other city uses of the site surface, including a planned playfield.

4.0 REFERENCES

AECOM; Golder Associates Ltd. (2014) *Water Supply Master Plan Update. Draft Final Report.* Prepared for the City of Guelph, Ontario, Canada, May 2014.

C3 Water and Gauley Associates Ltd. (2016) *2016 Water Efficiency Strategy Update, Version 5.0.* Prepared for the City of Guelph, Ontario, Canada, September 2016.

CH2M Hill (2009) *Guelph Wastewater Treatment Master Plan.* Prepared for the City of Guelph, Ontario, Canada, April 2009.

City of Guelph (2013) *Water Supply Master Plan.* http://guelph.ca/plans-and-strategies/water-supply-master-plan/ (accessed Feb 2017).

City of Guelph (2016) *Water Efficiency Strategy.* www.guelph.ca/wesu (accessed Feb 2017).

City of Guelph (2017) *About the Blue Built Home Program.* http://guelph.ca/living/environment/water/water-conservation/blue-built-home/about-the-blue-built-home-program/ (accessed Feb 2017).

City of San Diego (2016) *Completing our Water Cycle, Securing our Future.* www.sandiego.gov/sites/default/files/pure_water_san_diego_fact_sheet_9-15-16_1.pdf (accessed June 2017).

Elkington, J. (1997) *Cannibals with Forks: The Triple Bottom Line of 21st Century Business*; Capstone Publishing: Oxford.

Environment Canada (2016) *Residential Water Use in Canada.* https://www.ec.gc.ca/indicateurs-indicators/default.asp?lang=en&n=7E808512-1 (accessed Feb 2017).

Genivar, Inc. (2012) *Technical Memorandum—Effluent Reuse System—"Purple Pipe System"*. Prepared for the City of Guelph: York Trunk Sewer and Paisley-Clythe Feedermain Municipal Class Environmental Assessment, June 2012.

Ginzel, L. E.; Kramer, R. M.; Sutton, R. I. (1992) Organizational Impression Management as a Reciprocal Influence Process: The Neglected Role of the Organizational Audience. In *Research in Organizational Behavior*, Cummings, L. L., Staw, B. M., Eds.; **14**: 227–266; JAI Press: Greenwich, Connecticut.

GoldSim (2017) *Monte Carlo Simulation*. http://www.goldsim.com/Web/Introduction/Probabilistic/MonteCarlo/ (accessed Feb 2017).

Institute for Sustainable Infrastructure (2015) *Envision Rating System for Sustainable Infrastructure*. http://research.gsd.harvard.edu/zofnass/files/2015/06/Envision-Manual_2015_red.pdf (accessed June 2017).

Institute for Sustainable Infrastructure (2017) *Envision's Critical Role in Infrastructure Sustainability*. http://sustainableinfrastructure.org/training-material/ (accessed Feb 2017).

Lumina Decision Systems (2017) *Monte Carlo Simulation and Risk Analysis*. http://www.lumina.com/technology/monte-carlo-simulation-software/ (accessed Feb 2017).

Mind Tools (2017) *Decision Matrix Analysis*. https://www.mindtools.com/pages/article/newTED_03.htm (accessed Feb 2017).

National Research Council (2014) *Sustainability Concepts in Decision-Making: Tools and Approaches for the US Environmental Protection Agency*. https://www.nap.edu/read/18949/chapter/5#31 (accessed Feb 2017).

Orange County Water District (undated) *Frequently Asked Questions*. http://www.ocwd.com/gwrs/frequently-asked-questions/ (accessed Feb 2017).

Perrow, C. (1970) *Organizational Analysis: A Sociological View*; Wadsworth: Belmont, California.

Stantec Consulting Ltd. (2016) Environmental Impact Study of Residential Water Softeners for the City of Guelph and Region of Waterloo. Prepared for the Region of Waterloo and the City of Guelph, Ontario, Canada, December 2016.

U.S. Environmental Protection Agency (undated) *Use of Monte Carlo Simulation in Risk Assessments*. https://www.epa.gov/risk/use-monte-carlo-simulation-risk-assessments (accessed Feb 2017).

U.S. Geological Survey (2016) *The USGS Water Science School*. https://water.usgs.gov/edu/qa-home-percapita.html (accessed Feb 2017).

Water Research Foundation (estimated publication 2017) *Framework for Evaluating Alternative Water Supplies: Balancing Cost with Reliability, Resilience and Sustainability*, Project 4615; Water Research Foundation: Denver, Colorado.

4

Regulations and Risk Assessment

Caroline Russell, Ph.D., P.E., BCEE; Eva Steinle-Darling, Ph.D., P.E.;
Daniela Castañeda, P.E.; Bob Hultquist, P.E.; and Craig L. Riley, P.E.

Reuse of recycled water, defined in this book as "municipal wastewater that has been treated to meet specific water quality criteria with the intent of being used for beneficial purposes", is practiced in many countries around the world, particularly for agricultural and other nonpotable uses (U.S. EPA, 2012). The World Health Organization (WHO) has established guidelines for global water reuse (WHO, 2006). Several countries have developed country-specific guidelines (e.g., Australia) or regulations (e.g., India, Spain, Vietnam) for the use of recycled water.

In the United States, no federal regulations specific to reuse of recycled water have been promulgated. However, regulations established under the Clean Water Act (CWA) and Safe Drinking Water Act (SDWA) apply to aspects of water reuse projects. For example, discharge of recycled water to "waters of the state" before downstream reuse is regulated under the National Pollutant Discharge Elimination System (NPDES) set forth under the CWA. Industrial discharges to the sewershed that can affect the quality of water resource recovery facility (WRRF) effluent are regulated under the National Pretreatment Program, also established under the CWA. The Underground Injection Control (UIC) regulations found in Title 40 of the Code of Federal Regulations (CFR) set forth requirements for injection of water, including recycled water, to an aquifer and can affect water reuse projects for groundwater replenishment, to prevent seawater intrusion, or for indirect potable reuse (IPR). Any water intended for potable distribution, including purified water from an indirect or direct potable reuse (DPR) facility, must meet all National Primary Drinking Water Regulations (NPDWRs) established under the SDWA.

The U.S. Environmental Protection Agency (U.S. EPA) has established guidelines for water reuse, to which states can refer to establish criteria or standards based on the reuse application (U.S. EPA, 2012). As outlined in the guidelines (U.S. EPA, 2012), 43 states have established rules, regulations, or guidelines for agricultural reuse to irrigate processed food crops and non-food crops. Nine states have established regulations or guidelines for IPR including California (CCR, 2015), Florida (FDEP, 2014), Virginia (VDEQ, 2014), and Washington (WSDE, 1997). Texas has regulated two DPR facilities on a case-by-case basis and the Texas Water Development Board funded the development of *Final Report: Direct Potable Reuse Resource Document* (Alan Plummer Associates, 2015). U.S. EPA (2012) summarizes state reuse regulations and guidelines for different reuse applications.

Regulations, where they exist for reuse of treated wastewater, depend on the end use of the recycled water and focus primarily on public health and environmental protection. Table 4.1 lists categories of reuse and general corresponding risk and regulatory considerations. Nonpotable reuse applications include, but are not limited to, agricultural irrigation, landscape irrigation (e.g., golf courses, recreational fields), industrial uses (e.g., cooling towers), in-building uses (e.g., for flushing toilets), and impoundments. Nonpotable reuse also includes intentional beneficial reuse for habitat restoration (e.g., wetlands, streamflow augmentation). Recycled water discharged to a surface waterbody or aquifer that is later withdrawn intentionally as a drinking water supply constitutes IPR. *Direct potable reuse* is defined as

TABLE 4.1 Regulatory and risk considerations based on reuse application.

Recycled water use	Risks	Regulatory approach to control the risks
Agricultural irrigation	• Public exposure to pathogens on food crops • Farm worker exposure to pathogens • Public concern regarding plant uptake of contaminants	• Treatment and bacteriological requirements are set commensurate with the consumer exposure considering the type of irrigation (surface, drip, or spray) and the nature of the food crop (eaten raw or processed)
Landscape irrigation	• Public exposure to pathogens • Cross-connections	• Treatment and bacteriological requirements commensurate with the exposure (e.g., playground or median strip) • Backflow control, pipe labeling/color distinction, pressurization
Industrial uses	• Worker exposure to pathogens • Cross-connections	• Treatment and bacteriological requirements commensurate with the exposure associated with the use • Internal and meter backflow control, pipe labeling/color distinction, pressurization
In-building uses	• Public exposure to pathogens • Cross-connections	• Treatment and bacteriological requirements commensurate with the exposure associated with the use(s) • Internal and meter backflow control, pipe labeling/color distinction, pressurization
Habitat restoration (e.g., wetlands, streamflow augmentation)	• Public exposure to pathogens • Adverse effect on aquatic life/ecosystem if site-specific habitat requirements are not adequately addressed	• Treatment and bacteriological and chemical requirements commensurate with the exposure associated with the use • Discharge limits to preserve the beneficial uses of receiving water
IPR and DPR	Public exposure to • Regulated chemicals • Pathogens • Unregulated organic chemicals • Contamination of the aquifer or surface water for IPR	• Compliance with maximum contaminant levels • Organism log reduction targets to meet a tolerable risk of infection • Multiple treatment barriers for microbial contaminants • Treatment to control diverse trace organic chemicals • Discharge limits to preserve the beneficial uses of receiving water

the introduction of recycled water that has undergone advanced treatment either directly to a drinking water treatment facility or to the distribution system. The absence of an environmental buffer distinguishes DPR from IPR. Because each reuse application results in different environmental effects and human exposure, corresponding regulations and risk factors vary depending on the application.

This chapter provides guidance for planners and managers from either the public or private sector to identify the applicable regulatory requirements and risk management considerations for their site-specific reuse application. The chapter is structured to present information on the topics related to regulatory requirements and risk management associated with different water reuse projects.

For each reuse application outlined in Sections 5.0 through 7.0, general regulatory considerations are provided along with recommended steps for the reader to identify the requirements specific to their project. Other legal and risk management considerations are also provided. A separate section is not included for intentional beneficial reuse for habitat restoration. Readers are encouraged to review *Guidelines for Water Reuse* (U.S. EPA, 2012) as a starting point to assess applicable state regulations and risks associated with this and other reuse categories that may not be explicitly covered within this chapter.

1.0 STEPS TO IDENTIFY REGULATORY AND RISK CONSIDERATIONS

Figure 4.1 outlines the general steps for planners and managers to follow to identify the regulations governing a given reuse project. Step 1 is to define the type of reuse application because regulations and risks vary depending on the type of reuse project. Step 2 is to identify the general regulatory framework for the reuse project in the jurisdiction of interest. The purpose of this step is not necessarily to conduct a comprehensive evaluation of applicable regulations, but to gather sufficient familiarity to facilitate a discussion with state agencies to confirm state-specific requirements. Step 2 may include review of Sections 5.0 through 8.0 of this chapter, as applicable, coupled with review of U.S. EPA's (2012) *Guidelines for Water Reuse*, the relevant state agency websites, and/or a brief scan of state regulatory codes (e.g., Arizona Code of Regulations). Early in the project planning process, water rights and contractual agreements also need to be reviewed (Step 3) to identify the following: (1) the quantity of recycled water available for the reuse application under consideration and (2) any implications of

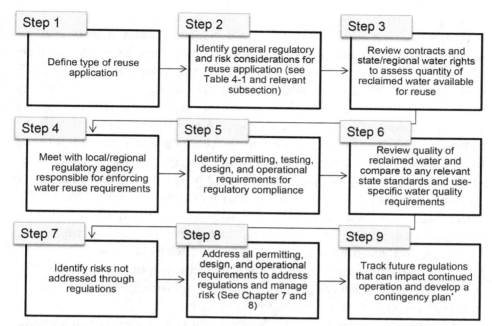

FIGURE 4.1   Approach for identifying and addressing regulatory requirements and managing risk.

*Contingency plan should include identification of a backup water supply for the end use (if needed) in the event regulations are promulgated that prevent continued operation.

discharging the recycled water to a surface or groundwater before extraction for reuse. Section 2 provides information on water rights considerations.

After a preliminary assessment of applicable regulations and water rights that could affect the magnitude and concept of the reuse project, planners/managers are encouraged to meet with the reuse regulatory agency for the project's jurisdiction (Step 4) to gauge any fatal flaws from a regulatory perspective and to identify specific permitting steps, testing requirements, and design requirements for regulatory compliance. Note that this meeting could also occur earlier in the process, with additional meetings as needed to confirm the planner/manager's understanding of applicable regulations. Additional details regarding permitting, testing, design, and operational requirements (Step 5) can be compiled after the meeting with regulators based on review of regulations and any references provided by the regulators. At this stage, a thorough review of the relevant state-specific rules is recommended, not only for compliance, but also to consider and address the basis for the rule (e.g., putting in measures not just for compliance, but to provide adequate health protection that was the motivation for the rule).

Step 6 is to review the quality of the recycled water for comparison with any relevant state standards and use-specific water quality requirements. This step facilitates identification of any treatment needed to meet regulations, and aligns closely with the approach outlined in Chapter 7. Presentation of preliminary water quality data to state regulators early in the planning process can also be a useful tool to garner additional agency support and guidance for the reuse project. A review of water quality is also critical to identify any constituents in the recycled water that would need to be removed to meet requirements for the specific reuse application (e.g., ammonia removal is generally recommended before use for toilet flushing to minimize aesthetic concerns associated with nitrifying biofilm growth). The water quality review also facilitates evaluation of risks under Step 7 that are not tied specifically to regulations, but to other public health, application efficiency, and environmental risks associated with the reuse application. Section 4 introduces the hazard analysis and critical control point (HACCP) approach that has been adopted from the food industry for risk management in reuse projects, in particular, for potable reuse.

As illustrated in Figure 4.1, it is important to continually track regulations and other legal requirements affecting reuse projects because the legal framework governing reuse is growing and changing as the scientific basis for regulation improves and more utilities and private companies are looking to recycled water to meet their future water needs.

2.0   WATER RIGHTS

Water rights are governed by state and tribal statute, regulations, and case law. For some river systems, such as the Columbia, Rio Grande, and Colorado Rivers, which are large, multistate, and international watercourses, water allocations are affected by specific treaties (Columbia, Rio Grande, and Colorado River Compacts, respectively). For historical and climatological reasons, water rights in each state are governed by one of two basic doctrines: riparian or appropriative. Under the riparian water rights doctrine, the right to use water is derived from land ownership adjacent to (or above, in the case of groundwater) a watercourse. The holder of this kind of water right is entitled to use water as long as that use does not interfere with the other riparian user's ability to use the water. This type of water rights system focuses on maintaining sufficient in-stream flows for all users and does not allow or account for storage of water. Riparian water rights systems are dominant in areas where water has not historically been scarce, such as the eastern United States. Under an appropriative doctrine, water rights are established from the date of first beneficial use or priority date, under the so-called "first in time, first in right" or "priority" principle. At statehood, each state was granted ownership of all of the water within its boundaries. State regulatory agencies assign water rights based on demonstration of four principles: the water is put to beneficial use, water is available, no existing water rights are impaired, and the water right is in the public interest. The ability to withdraw water is dependent on seniority, which depends on the date when the water right was issued. In times of shortage, those with the most "junior" rights see their water rights cut off first. Under appropriated water law, water rights are considered to be real property and, as such, may be sold, leased, or transferred the same as a parcel of land. Appropriative water rights systems are dominant in the western United States in areas that have been historically water-limited.

When it comes to implementing water reuse projects, the main question affecting the ability to reuse some or all of the recycled water produced is, typically, who has legal access to an existing effluent discharge. Beneficial reuse of that water necessarily results in less water being returned to the watercourse. The question is if, when, and how much of that effluent becomes a part of the flow in the receiving water that is subject to appropriation. In some states, water rights are granted "to extinction", whereas, in others, rights are granted with a provision that requires a certain quantity of "return flows" to state waters to ensure minimum streamflow levels to maintain and support riparian habitats.

It is in the question of "Who owns the effluent?" that nuances in water rights law can make an enormous difference in an entity's ability to reuse water beneficially. For example, in Texas, water rights are generally granted to extinction. This means effluent that is put to immediate or "direct" beneficial reuse requires no additional water right. However, while there is no requirement to do so, once discharged into a watercourse, the effluent becomes subject to re-appropriation. A discharger may apply for a right to reuse some of that water downstream, a practice referred to as *indirect reuse*, by submitting a request to use the "bed and banks" of the state's watercourse to transport their water to a designated point of diversion, where the entity then receives, by definition, a junior right to that water. In addition, the volume that can be diverted is reduced by calculation of in-stream losses, and, in the case of over-allocated watercourses, a right may not be granted at all (Rochelle, 2008). This inevitable loss of both volume and priority provides an uncommonly strong incentive to pursue more DPR projects over the more common IPR alternative. Similarly, Washington and Florida grant to the effluent producer the "exclusive right" to any recycled water produced, effectively exempting it from the water rights process. However, once discharged to a natural waterbody, this water becomes subject to reappropriation.

An overview of the water rights doctrine and relevant regulatory requirements for a number of states are listed in Table 4.2. Recommended steps to assess applicable water rights considerations for a proposed potable reuse project are as follows:

1. Review any contractual agreements for the recycled water supply;
2. Identify the state/regional agency responsible for surface and/or groundwater rights allocation within the jurisdiction; and
3. Assess any effects of current contractual agreements and/or water rights law on the recycled water available for the proposed reuse activity, along with how any proposed surface or groundwater discharge before reuse could affect subsequent rights for extraction.

3.0 WATER QUALITY

General water quality parameters of importance for the range or reuse applications are illustrated in Table 4.3. *Water Reuse Issues, Technologies, and Applications* (Asano et al., 2007) provides summary information on ranges of concentrations that may be observed for some of the parameters highlighted in Table 4.3 in raw and treated wastewater, depending on the treatment achieved at the WRRF. Data on trace organic and pharmaceutical concentrations measured in raw and treated wastewater have been published elsewhere (Baronti et al., 2000; Falas et al., 2016; Salveson et al., 2012; Sedlak and Kavanaugh, 2006). Because the concentrations of different microbial and chemical constituents in recycled water vary depending on sewershed characteristics (particularly if industrial discharges are permitted as part of a pretreatment program) and on the treatment provided at the WRRF, site-specific characterization of the recycled water via a robust sampling program is a recommended (and, in some states, required) step for potable reuse applications.

The parameters of key interest, and those that should be sampled as an initial planning step for a reuse project, depend on the reuse application and geographic location. For example, helminths present a concern during agricultural and land irrigation, primarily in developing countries where nematode and other waterborne diseases transmitted through dermal exposure are of greater concern. Metals and inorganics that can scale on process components can hinder operational efficiency and long-term operability in cooling water and other industrial reuse applications. High concentrations of salts pose challenges for agricultural irrigation. Table 4.3 lists some of the use-specific considerations. Additional water quality considerations

TABLE 4.2 Water rights provisions related to reuse in selected states.

State	Water rights paradigm	Key water rights provisions related to reuse
California	Appropriative	California Water Code §1211(a) requires approval by the California Water Resources Control Board of downstream impacts of diverting flows for reuse.
Colorado	Appropriative	Water reuse is covered by the "Blue River Decree" and other rulings in State Water Court. Only nontributary water (e.g., interbasin diversion or non-tributary groundwater) may be reused.
Florida	Neither (see Florida Statue 373)	Florida water law does not restrict the ability to reuse effluent from a water rights perspective. In fact, for effluent dischargers within water resource caution areas, Florida law requires the implementation of reuse if feasible (Florida Administrative Code 62-40.416 (2)).
Illinois	More riparian	Illinois Department of Natural Resources allocates water use from Lake Michigan based on a set of priorities that include proximity to the lake. Illinois does not restrict the ability to reuse effluent from a water rights perspective.
Nevada	Appropriative	Nevada Department of Water Resources reviews reuse applications to determine whether diversion of historical discharges will impair the rights of downstream users. Rights to the Colorado River are governed by the Colorado River Compact. Because rights to the Colorado are made as consumptive use (net) allocations, users can obtain "return flow credits" for all effluent discharges. This incentivizes conservation and non-consumptive uses, such as IPR over consumptive uses such as irrigation.
New Jersey	Riparian	New Jersey does not restrict the ability to reuse effluent from a water rights perspective.
Oregon	Appropriative	Dischargers who want to reuse their effluent must submit a "Municipal Reclaimed Water Registration" form with the Oregon Water Resources Department (OWRD), who may limit reuse of recycled water "if the municipality has discharged the recycled water into a stream for five or more years, and the discharge represented more than 50 percent of the total average annual flow of the stream" (OWRD, undated).

(continued)

TABLE 4.2 Water rights provisions related to reuse in selected states (*Continued*).

State	Water rights paradigm	Key water rights provisions related to reuse
Texas	Appropriative	Water rights are granted by the Texas Commission on Environmental Quality (TCEQ) and can be used extinction. This means direct reuse needs no additional right (Texas Water Code §11.064(c)). However, once discharged, effluent becomes "surplus water" that is subject to re-appropriation. Rights to water from indirect reuse can be granted through a bed and banks permit with junior water rights at the point of downstream diversion (Texas Water Code §11.042).
Utah	Appropriative	Water rights in Utah are issued as diversion rights that often include a requirement to provide a certain amount of return flow. An application for reuse to the State Engineer's Office must include a calculation of the additional anticipated consumptive use or "depletion" of that water, which must be consistent with the original water right (Utah Code Ann. §73-3c-302(2)a-c).
Washington	Appropriative	Recycled water production, distribution, use, storage, and recovery exempt from water rights process; state statute protects downstream water rights from impairment by upstream diversions of discharges for reuse without mitigation or compensation (Wash Rev. Code §90.46.130(1)). Irrigation and potable supplies artificially stored underground require an additional storage permit for recovery in addition to the original water right.

and regulations specific to different reuse applications are provided in Sections 5.0 through 7.0.

4.0 RISK IDENTIFICATION AND MANAGEMENT

Compliance with regulations addresses the health and environmental risks officially recognized by the jurisdiction. Additional risks that should be considered include the following:

- Regulatory noncompliance resulting from upsets in source water quality, operations, and/or treatment performance—the HACCP framework is emerging as the water reuse industry standard for putting in place process controls that will detect and correct deviations in

TABLE 4.3 Water quality parameters affecting regulatory compliance and risks in reuse applications.

Water Quality Parameters				Regulatory and Risk Considerations
Pathogens				• Risk of infection from dermal contact (agriculture or landscape irrigation), inhalation (aerosols during spray irrigation or use in cooling towers), or oral consumption
Viruses	Cryptosporidium & Giardia	Bacteria	Helminths	• State requirements for log removal of pathogens for potable reuse applications
				• State requirements for total and/or fecal coliform bacteria for non-potable reuse applications
Inorganic Chemicals				• Primary MCLs or potable reuse or injection into groundwater
Sodium, Chloride, Total Dissolved Solids	Ammonia, Nitrate, Perchlorate, Phosphate	Metals (e.g., arsenic, barium, chromium, copper, iron, lead, manganese, silica, strontium)	Radionuclides	• Metal scalant impacts to membrane treatment processes (advanced treatment for potable or industrial reuse), and production applications (industrial reuse)
				• Ammonia leading to aesthetic concerns from nitrifying biofilm growth and process control considerations for advanced treatment
				• TDS impacts to aesthetic considerations for potable reuse
Organic Chemicals				• Primary MCLs for potable reuse or injection into groundwater
Regulated volatile organic compounds	Semi-volatile organic compounds			• Impacts to organic certification for agricultural reuse
				• Formation of disinfection by-products (DBPs) from reaction of disinfectants with bulk organic matter
Chemicals of Emerging Concern				• Public perceptions and concerns (e.g., associated with exposure to endocrine disrupting compounds)
Pharmaceuticals / Personal Care Products	Endocrine Disrupting Compounds	Perfluorinated Compounds	Unregulated herbicides / pesticides	Other anthropogenic compounds	• Health risk associated with perfluorinated compounds
				• Impacts to organic certification in agricultural reuse applications for constituents on the National List of the USDA National Organic Program
Disinfection By-Products				• Primary MCLs for regulated DBPs for potable reuse or injection into groundwater
Regulated (bromate, chlorite, TTHM, HAA5)	Unregulated (NDMA, chlorate, brominated and iodinated DBPs)			• Potential health risks and future regulatory implications for unregulated DBPs; NDMA is of particular concern for potable reuse applications given the 10 ng/L California Notification Level (NL) and formation during ozonation of reclaimed water
General / Physical Parameters				• Impacts to any downstream water treatment processes
Particulates (measured as total suspended solids, turbidity)	Alkalinity	pH, dissolved oxygen (DO) & temperature	Total organic carbon, peCOD	• Temperature impacts to industrial and urban non-potable reuse applications
				• Excessive particulates can lead to clogging in appurtenances in non-potable reuse applications and reduction in treatment efficiency and/or operational impacts for advanced treatment for potable reuse applications

process performance and quality at the earliest possible opportunity (Halliwell et al., 2014). Chapter 8 provides guidelines on monitoring and control for reuse projects.

- Emerging health or environmental concerns for which no regulations or standards currently exist—because reuse science and regulation are developing, planners/managers are advised to check with the regulator to see if there are emerging concerns for which additional regulation is being considered. Depending on the reuse application, additional safeguards may be incorporated to prevent exposure to unregulated contaminants of emerging concern. For example, diverse treatment barriers capable of reducing concentrations of chemicals of emerging concern to low-part-per-trillion concentrations are recommended in potable reuse projects despite the absence of corresponding regulatory statues or clear epidemiology or toxicology data demonstrating a health risk.

- Recycled water quality or quantity that is within statutory requirements, but that affects the reuse application operational efficiency and long-term viability—recycled water customers should be consulted to see if they are concerned about additional risks or have water quality requirements not addressed by the regulations. As an example, industrial customers may be concerned with nutrients that could lead to biofilm growth in cooling towers, whereas agricultural reuse customers may prefer to receive recycled water with higher nutrient concentrations. West Basin Municipal Water District's Water Recycling Facility in Carson, California, provides a good example of how the statutory and nonregulated water quality requirements are met by different levels of treatment depending on the reuse application.

The project management should become familiar with the risks inherent in the planned uses and the considerations involved in managing the risk using available references (e.g., Asano et al., 2007; Halliwell et al., 2014). The stakes are high, particularly for potable reuse, foremost to prevent and mitigate any potential health risk, but also because any perceived risk can jeopardize public confidence in the reuse program. Risk management considerations associated with specific reuse applications are discussed further in the following sections.

5.0 AGRICULTURAL IRRIGATION

Globally, agriculture accounts for approximately 70% of freshwater withdrawals (UN Water, 2014) and, in the United States, the agriculture sector accounts for 80% of water withdrawals (USDA, 2012). Use of recycled water for agricultural irrigation can significantly reduce the freshwater withdrawal rate. In Florida, recycled water has been used for agricultural irrigation for more than 40 years (Southwest Florida Water Management District, 2009). Use of recycled water in Monterey, California, for the last 18 years has been attributed to more sustainable use of groundwater, reduced saltwater intrusion, and the production of high quality crops with no associated incidence of adverse health effects from crop consumption or worker exposure (Crook, 2004; Engineering-Science, 1987).

5.1   Current Regulatory Framework

U.S. EPA has provided guidelines for use of recycled water for agricultural irrigation and has encouraged states to develop additional guidelines or standards (U.S. EPA, 2012). Twenty-seven states have developed regulations, standards, or guidelines for the use of recycled water to irrigate unprocessed food crops (i.e., crops consumed raw without processing), and 43 states have developed regulations or guidelines for irrigation of processed food crops and nonfood crops (U.S. EPA, 2012), although the level of regulation varies widely between states. Most states have established tiered requirements for water quality based on the type of agricultural use. For example, in California, the total coliform standard for spray-irrigation of food crops is 2.2 colonies/100 mL (most probable number [MPN]/100 mL), whereas the standard for pasture irrigation for milking animals is 23 MPN/100 mL (SWRCB, 2016). Arizona has established five classes of recycled water under Arizona Administrative Code (AAC) Title 18, Chapter 11, Article 3, with corresponding minimum treatment requirements and water quality criteria. In Florida, direct contact (spray) irrigation of edible crops that will not be peeled, skinned, cooked, or thermally processed before consumption is not allowed except for tobacco and citrus.

Table 4.4 presents a summary of the treatment requirements for food crop irrigation for the states of California, Arizona, Florida, and Texas. All four states include criteria for either total or fecal coliform, reflecting requirements for pathogen reduction for public health protection.

Notably, none of the states includes requirements for total dissolved solids (TDS), which does not represent a public health issue. Research shows that recycled water, while exhibiting higher TDS than some conventional supplies, can be applied to crops without hindering production.

A National Academy of Sciences study conducted by the National Research Council (1996) concluded that these and other state standards along with "site restrictions and generally good system reliability have insured that food crops thus produced do not present a greater risk to the

TABLE 4.4   Agricultural water reuse regulations for food crops.[a]

Reuse requirements	Arizona	California	Florida	Texas
Regulatory enforcement agency	ADEQ	SWRCB	FDEP	TCEQ
Regulatory statute	AAC Title 18	CCR Title 22	FAC Chapter 62-610	30 TAC 210 Subchapter C
Categorization	Class A	Disinfected Tertiary	Not specified	Type I
Treatment requirements[b]	Secondary treatment, filtration, disinfection	Oxidized, coagulated, filtered, disinfected	Secondary treatment, filtration, disinfection	Not specified
Water quality requirements:				
5-day carbonaceous biochemical oxygen demand ($CBOD_5$)	Not specified	Not specified	60 mg/L maximum; 20 mg/L annual average[c]	5 mg/L
TSS	Not specified	Not specified	5 mg/L maximum	Not specified
Turbidity	2 NTU daily average 5 NTU maximum	2 NTU average for media filters[d]	Case by case; continuous online monitoring required[e]	3 NTU
Bacterial indicators	Fecal coliform: none detectable in last 4 of 7 samples	Total coliform: 2.2 MPN/100 mL 7-day median	Fecal coliform: 75% of samples below detection[f]	Fecal coliform
Pathogens	No detectable enteric virus in 4 of last 7 monthly samples	5-log (99.999%) virus inactivation goal	Protozoa sampling once per 2-year period for facilities ≥0.044 m³/s (1 mgd)	Not specified
Other	If nitrogen >10 mg/L, special requirements may be mandated to protect groundwater	Not specified	Not specified[g]	Not specified

[a]ADEQ = Arizona Department of Environmental Quality; SWRCB = State Water Resources Control Board; FDEP = Florida Department of Environmental Protection; and FAC = Florida Administrative Code.
[b]California and Florida further specify chlorine disinfection requirements; UV dose if UV disinfection is used should adhere to the National Water Research Institute UV guidelines.
[c]Regulations also specify 30-mg/L monthly average and 45-mg/L weekly average $CBOD_5$.
[d]California regulations specify 10 NTU maximum for media filters, 0.2 NTU average and 0.5 NTU maximum for membrane filters.
[e]Generally, 2 to 2.5 NTU is required.
[f]New Food Safety Modernization Act requirements for *Escherichia coli*.
[g]Florida regulations specify total nitrogen requirements for some reuse applications, such as rapid infiltration basins, which could apply depending on the approach for agricultural reuse.

consumer than do crops irrigated from conventional sources". A more recent report developed by the National Water Research Institute (NWRI) blue-ribbon panel evaluated the adequacy of regulations in the State of California on the use of recycled water for irrigation of food crops (NWRI, 2012). The panel concluded the following:

> The bottom line is that the median annualized risk estimates for infection are consistent with previous estimates relied on by CDPH to develop the Water Recycling Criteria [CDPH considers a 1 in 10,000 (i.e., 1×10^{-4}) mean risk of infection to be an acceptable risk from exposure to treated wastewater effluent (Yamamoto, 2010).] and, as discussed below, provided the Panel with additional evidence to confirm the conclusion that current agricultural practices that are consistent with the criteria do not measurably increase public health risk, and that modifying the standards to make them more restrictive will not measurably improve *public health* (NWRI, 2012).

Additional regulatory requirements for agricultural reuse can include stipulations for irrigation method (e.g., spray, surface, subsurface) to prevent exposure to aerosols, worker safety requirements, and use of appropriate setbacks between irrigated areas and potable water connections. Planners and managers considering agricultural reuse programs in their state/region should review their state-specific regulations to identify specific treatment, water quality, and other requirements depending on the particular application (e.g., nonprocessed food crops, processed food crops, or nonfood crops). Regulations for water reuse projects typically require a substantial demonstration of water quality at the end point of treatment, detailed documentation of each use site, and periodic inspections of each use site. In some cases, trained site supervisors are required to be designated for each site to ensure cross-connection control and compliance with other water recycling rules. U.S. EPA's (2012) *Guidelines for Water Reuse* and other reports referenced at the end of this chapter provide guidance on water quality considerations for agricultural reuse applications.

5.2   Other Legal Considerations

In addition to compliance with any state regulatory requirements, agricultural reuse applications will need to adhere to any water rights considerations and specific criteria for food production. Contractual agreements for recycled water supply may be executed for firm allocation of the water supply for agricultural production—sometimes with stipulation of seasonal allocations—and the duration of the agreement. If the recycled water is first discharged to a "water of the state", an NPDES permit is required.

While the corresponding WRRF likely already has an NPDES permit on file, a change in discharge location to facilitate agricultural reuse would necessitate a change in the existing permit or application for a new permit. To minimize regulatory paperwork for each separate use site, some agricultural reuse projects are negotiating umbrella permits with the agencies producing and distributing the recycled water to cover multiple use sites/individual farmers.

The U.S. Department of Agriculture (USDA) is responsible for establishing national standards for organic food production. The USDA organic regulations do not include irrigation and water quality requirements. However, the regulations stipulate that crops and soils may not be contaminated with prohibited substances on the National List (Title 7 Subtitle B, Chapter I, Subchapter M, Part 205 CFR). Producers should take precautions to ensure that irrigation water is not loaded with agricultural pesticides or other polluting chemicals (USDA, 2012). By interpretation, provided that recycled water does not include these substances, there are no restrictions to its use to irrigate organic crops. There are several examples of recycled water used to irrigate organic crops in the country, including organic farmland in California's Northern Monterey County irrigated with recycled water from the Monterey Wastewater Reclamation Study for Agriculture (Engineering-Science, 1987).

5.3   Risk Assessment and Management

Public health risks associated with agricultural irrigation with recycled water are low in the United States and in other developed countries, as evidenced by decades of application with no documented public health implications, and reflective of regulations, statutes, and/or guidelines that establish the necessary safeguards to avoid adverse health effects. In developing countries, where wastewater is often not treated to the same quality, health effects from crop irrigation with untreated or partially treated wastewater are more of a concern. Low-cost strategies to mitigate risks from reuse of partially treated wastewater in developing countries have been suggested by Scott et al. (2010) and Amoah et al. (2011). The WHO (2006) guidelines for water reuse are a resource for setting goals to minimize health risks associated with agricultural reuse worldwide.

The HACCP approach, while not historically adopted within the agricultural reuse sector, could be applied to further reduce risks, principally by incorporating monitoring at defined control points. The HACCP approach could be of specific interest for crops consumed raw and could provide assurance to the growers and the public that the reuse project is a full partner in food safety. Critical control points (CCPs) can be identified in each treatment process that provides barriers for pathogens or other constituents of concern.

Additional risks warranting considerations for agricultural reuse projects include the following: (1) the effect on long-term soil productivity and crop production; (2) potential exposure to trace contaminants; and (3) public concern related to reuse practices, with potential effects on product sales. Certain crops can be sensitive to the salinity of reuse water. The sodium absorption ratio in the recycled water should be evaluated and discussed with growers to identify any potential issues affecting long-term soil viability for agricultural production.

Plant uptake of chemicals of emerging concern (CECs) has been raised as a potential public health consideration. Two recently published meta-analyses by Prosser and Sibley (2015) and Wu et al. (2015) provide an up-to-date overview of the plant uptake of pharmaceuticals and personal care products (PPCPs) in connection to wastewater irrigation and biosolids in agriculture. These analyses provide a significant contribution to the state of the science on the uptake of these compounds in edible crop; however, some of the studies included in the analyses exposed the plants and soil to CECs and PPCPs at concentrations well above these compounds' concentrations present in wastewater. Thus, the results are not a real representation of what is occurring under field conditions and may overestimate the real uptake or concentrations in plant tissues and the soil. A recent study evaluated the concentrations of 19 commonly occurring PPCPs, including caffeine, meprobamate, primidone, DEET, carbamazepine, dilantin, naproxen, and triclosan in eight vegetables (commonly used in salads and often consumed raw) irrigated until harvest with filtered and disinfected secondary effluent with and without a fortified dose of PPCPs. Results from the study showed that accumulation of the PPCPs was limited under field conditions, translating to dietary exposure to PPCPs more than 3 orders of magnitude smaller than a single medical dose for one compound through daily consumption of the vegetables irrigated with wastewater that was *fortified* with PPCPs (Wu et al., 2014).

Despite a long history of successful agricultural reuse of recycled water, the press and elected officials do revisit the practice in response to reports of a foodborne disease outbreak. Therefore, proactive development of a communication and response strategy is recommended as a tool to respond to any associated public concerns. Furthermore, some wholesalers still do not accept produce irrigated with recycled water. Work still needs to be done to develop public relations campaigns to alleviate public concern and minimize risks associated with a reduction in sales from irrigation with recycled water. Chapter 6 provides guidelines for public communication and outreach related to different reuse programs.

6.0 URBAN REUSE

Along with agricultural irrigation, urban reuse is one of the most significant applications of recycled water. Urban reuse primarily includes irrigation for landscapes, golf courses, and recreational fields, although other in-building uses, such as toilet and urinal flushing, are becoming more prevalent. Urban reuse opportunities discussed in this chapter also include uses in industrial applications, such as cooling towers and snowmaking. The commonality between these nonpotable, typically urban reuse applications is the inclusion of a separate transmission system conveying the recycled water from the WRRF to customers for reuse. With increases in population density, urban reuse of recycled water is an important tool to help water utilities address water supply challenges (Garcia-Cuerva et al., 2016).

6.1 Current Regulatory Framework—Global, Federal, State

Recycled water intended for urban reuse in municipal locations is divided into two categories: "unrestricted", also known as "Type 1", which denotes uncontrolled public access and, therefore, has more stringent regulations and "restricted", also known as "Type 2", which signifies a level of control over public access and, consequently, minimizes or eliminates risks of exposure. Industrial reuse is commonly categorized as a separate application because water quality requirements can be significantly higher to facilitate the intended use.

A substantial number of states have developed reuse regulations and guidelines for urban and industrial applications, as summarized in *Guidelines for Water Reuse* (U.S. EPA, 2012), although the level of regulation varies extensively. Table 4.5 provides a summary of treatment and water quality requirements corresponding to the unrestricted category, which includes industrial uses, or restricted reuse applications in California and Florida, providing an example for the types of requirements that may be imposed by states. Planners and managers are encouraged to review *Guidelines for Water Reuse* (U.S. EPA, 2012) as well as state regulatory agency websites and codified rules as a starting point to assess other states' specific standards for unrestricted and restricted urban reuse applications.

As illustrated in Table 4.5 for California and Florida, treatment and water quality requirements are typically more stringent for unrestricted rather than restricted reuse applications because of the greater potential for human contact. Florida established water quality requirements for 5-day carbonaceous biochemical oxygen demand ($CBOD_5$) and total suspended solids (TSS), generally limiting the organic and particulate loading into the recycled water system. Particulates, as measured by TSS and turbidity, can

TABLE 4.5 Urban water reuse regulations.*

Reuse requirements	California		Florida	
Regulatory enforcement agency	SWRCB		FDEP	
Regulatory statute	CCR Title 22		FAC 62-610	
Type of use	Unrestricted or industrial	Restricted	Unrestricted or industrial	Restricted
Categorization	Disinfected tertiary	Disinfected secondary	Highly disinfected secondary	Disinfected secondary
Treatment requirements	Secondary treatment, oxidation, coagulation, filtration, disinfection	Secondary treatment, oxidation, disinfection	Secondary treatment, filtration, high-level disinfection	Secondary treatment, disinfection

Water quality requirements:

	California		Florida	
$CBOD_5$	Not specified	Not specified	$CBOD_5$: • 20 mg/L (annual average) • 30 mg/L (monthly average) • 45 mg/L (weekly average) • 60 mg/L (maximum)	Not specified
TSS	Not specified	Not specified	5 mg/L (maximum)	10 mg/L (maximum for subsurface applications)
Turbidity	• 2 NTU (average) for media filters • 10 NTU (maximum) for media filters • 0.2 NTU (average) for membrane filters • 0.5 NTU (maximum) for membrane filters	Not specified	Case by case (generally, 2 to 2.5 NTU) Florida requires continuous online monitoring of turbidity as indicator for TSS	Not specified

(continued)

TABLE 4.5   Urban water reuse regulations.* (*Continued*)

Reuse requirements	California		Florida	
Bacterial indicators	Total coliform: • 2.2/100 mL (7-day median) • 23/100 mL (not more than one sample exceeds this value in 30 days) • 240/100 mL (maximum)	Total coliform: • 23/100 mL (7-day median) • 240/100 mL (not more than one sample exceeds this value in 30 days)	Fecal coliform: • 75% of samples below detection • 25/100 mL (maximum)	Not specified
Pathogens	Not specified	Not specified	*Giardia* and *Cryptosporidium* sampling once each 2-year period for facilities ≥0.044 m³/s (1 mgd); once each 5-year period for facilities ≤0.044 m³/s (1 mgd)	Not specified

*SWRCB = State Water Resources Control Board; CCR = California Code of Regulations; FDEP = Florida Department of Environmental Protection; and FAC = Florida Administrative Code.

hinder disinfection efficacy and consequently result in compliance issues and human exposure risks. Elevated concentrations of organics (e.g., CBOD$_5$) can lead to biofilm growth and a loss in dissolved oxygen, which can cause issues in a recycled water system. Planners and managers will need to review and discuss state-specific regulations with their state regulators to ensure compliance depending on the type of urban or industrial reuse application being considered.

6.2   Other Legal Considerations

Some states require inclusion of precautionary signs, pipeline tagging, locked vaults, and other features to control public access and minimize exposure risk for recycled water systems. They have also established additional safety requirements for dual-piping systems, such as color-coding, to minimize potential risks from cross-connecting the drinking water system with the recycled water system. For most states, purple is the identification color for recycled water distribution pipelines, and, for some, it is the established color for in-building piping as well. Cross-connections are further addressed through standard safety practices, such as backflow prevention.

Additional information on cross-connection control can be found in the *Cross-Connection Control Manual* (U.S. EPA, 2003), *Risk-Based Framework for the Development of Public Health Guidance for Decentralized Non-Potable Water Systems* (WE&RF, 2017), and *Blueprint for Onsite Water Systems: A Step by Step Guide for Developing a Local Program to Manage Onsite Water Systems* (WRF, 2014).

Another common practice is separation (e.g., setbacks) and grade layout distances between distribution system pipes, whether potable, wastewater, or recycled water. For example, a potable water pipe will always be at a higher elevation than others to avoid contamination from leaks or breaks in the pipelines below. In some cases, cased piping is allowed when separation requirements are not feasible.

As states introduce or further develop existing recycled water regulations, particularly those applicable to decentralized building uses, additional work is required to appropriately update relatable or overlapping regulations/guidelines. A present challenge because of the lack of water reuse regulations at the federal level is the conflicting language within and among states between recycled water regulations, plumbing codes, and other state regulations (Yu et al., 2013). Planners and managers need to carefully review all applicable requirements and hold active communication with the regulatory enforcement agencies.

Planners and managers should also consider emerging data and studies that can affect future regulatory compliance (see Step 9, Figure 4.1) to facilitate development of an urban reuse system that accommodates potential future compliance requirements. For example, *Legionella* in recycled water was found to have inverse seasonal variations to potable water; the higher irrigation demands during the summer correlated with the lowest concentrations (Blanky et al., 2015). Therefore, if monitoring is to be required once annually, it should be conducted during the winter months when the highest counts are anticipated.

6.3   Risk Assessment and Management

Risk considerations for urban reuse of recycled water include the following: (1) adverse effects to the operation or activity for which the recycled water is used and (2) health risks not addressed through regulatory compliance. Risk considerations for some urban reuse activities, as well as public perception challenges, are discussed in the following subsections.

6.3.1 Landscape Irrigation

Salts can have an adverse effect in irrigation applications depending on the receiving flora. Golf courses are particularly sensitive to this because of the aesthetic expectations from their customers. Indeed, "purple gold" has become the label to characterize their relationship with recycled water use and the water quality ranges required for turf grass irrigation (Harivandi, 2011). Nutrients (e.g., nitrogen and phosphorus) can be beneficial for plant growth, but simultaneously pose concerns for runoff to waterbodies. If it is unfeasible to address certain constituent limits without significant financial investment in treatment, other potential solutions include diluting recycled water, optimizing soil drainage, and flushing excess salts through excess irrigation. Identifying plant species based on their tolerance for salts and metals and characterizing the local soil can also help select the most adequate turf grass to be used in a particular golf course. Alternating irrigation between recycled water and potable water is another option for domestic applications to lessen the water quality effects on plants (Pinto et al., 2010).

Access restriction and cautionary signage can be used to minimize risks both associated *and not associated* with the recycled water use. For example, pathogens present in urban soils irrigated with recycled water can sometimes be a result of factors beyond the water's microbial content, such as animal feces and food waste (Benami et al., 2013). Signs can help deter activities that could present risks that could be inadvertently attributed to recycled water use, presenting public concerns.

6.3.2 Decentralized Building Uses

Recycled water used for toilet flushing should be of suitable quality to prevent aesthetic concerns. Currently, unregulated pathogens, such as *Legionella*, can pose a concern and should be proactively addressed to prevent exposure. One recycled water system in southern California, which supplied un-nitrified recycled water to industrial customers for toilet flushing, irrigation, and cooling tower operation, began receiving complaints from a purple pipe customer resulting from nitrification in the recycled water pipes. Irrigation pipes were clogging and toilets developed an "orange-ish slime". It is essential to identify the required water quality in toilet flushing applications because it is the most permitted recycled water use in buildings across the United States (Yu et al., 2013). Guidance and safety instructions for management staff represent another important approach for proper building uses, plumbing maintenance, and cross-connection control.

6.3.3 Industrial Uses

Water quality constituents of concern for industrial applications partly depend on the end use. Commonly, particulates, minerals, organics, and nutrients present in the recycled water can be of concern because of the potential for clogging, scaling, and biofilm growth. Each industry can have unique or more stringent requirements than others, thereby adding a level of complexity to recycled water systems that need to meet various requirements and customer preferences. Centralized recycled water systems could meet varying demands through a "fit-for-purpose" model, with different levels of treatment provided depending on the end use (Chen et al., 2016). A portion of the effluent from each advanced treatment process can be routed to specific industrial users who accept that particular level of water quality. West Basin Municipal Water District's Water Recycling Plant provides an example of this type of system and can be a good reference for navigating regulations associated with treatment and monitoring requirements for the different end uses.

6.3.4 Public Perceptions

Public concerns regarding recycled water use must be considered for each type of urban and industrial reuse application. Public outreach plans should be developed proactively to curtail public concerns regarding urban reuse applications. Monitoring can support public outreach, demonstrating that the recycled water is routinely assessed to evaluate exposure risks and confirm regulatory compliance.

7.0 INDIRECT AND DIRECT POTABLE REUSE

De facto potable reuse has been occurring for centuries. In the 20th century, the first intentional potable reuse projects were implemented, with the Goreangab Water Reclamation Plant in Windhoek, Namibia, coming online in 1968 as the first DPR facility in the world; simultaneously, various IPR facilities were constructed in Australia, Singapore, Israel, the United States, and elsewhere. Table 4.6 lists some of the potable reuse projects implemented in the United States, adapted from a database compiled by the WateReuse Association and the Water Environment & Reuse Foundation (2017). The Windhoek, Namibia, facility is also included. These projects and others laid the groundwork for developing regulatory guidelines for permitting of the facilities and for identifying and managing risks associated with the continuum of potable reuse applications.

Regulations, guidelines, and risk management considerations for potable reuse reflect the need for a high degree of treatment, monitoring, and process control relative to nonpotable uses because of the much higher public exposure to the water (routine ingestion, dermal contact, and inhalation while bathing). Both IPR and DPR regulatory and risk management requirements are discussed in this section, reflecting the fact that the projects fall along a continuum, with the key difference between IPR and DPR being the

TABLE 4.6   Characteristics of select potable reuse case studies (adapted from WRA and WE&RF, 2017).*

Facility location	Facility name/flow[b]	Date operation started	Blending location[d]	Permitting agency
Los Angeles County Sanitation District, California	Montebello Forebay groundwater recharge project (1.93 m³/s [44 mgd])	1962	Groundwater recharge via soil-aquifer treatment (IPR)	California SWRCB
Windhoek, Namibia	New Goreangab Water Reclamation Plant (0.24 m³/s [5.5 mgd])[c]	Original facility in 1968; new facility in 2002	Potable distribution system (DPR)	City of Windhoek
Upper Occoquan Service Authority, Virginia	Regional Water Reclamation Plant (2.63 m³/s [60 mgd])	1978	Surface water augmentation (IPR)	Virginia Department of Health
El Paso, Texas	Fred Hervey Water Reclamation Facility (0.55 m³/s [12.5 mgd])	1985	Groundwater recharge via infiltration basins (IPR)	TCEQ
West Basin Municipal Water District, California	Water recycling facility (0.55 m³/s [12.5 mgd])	1993	Groundwater recharge via direct injection (IPR)	California SWRCB
Scottsdale, Arizona	Scottsdale Water Campus (0.61 m³/s [14 mgd])	1999	Groundwater recharge via direct injection (IPR)	ADEQ
Gwinnett County, Georgia	F. Wayne Hill Water Resources Center (2.37 m³/s [54 mgd])	1999	Surface water augmentation (IPR)	GA DNR EPD
Orange County, California	Groundwater replenishment systems (3.07 m³/s [70 mgd])	2008	Groundwater recharge via direct injection and spreading basins (IPR)	California SWRCB
Wichita Falls, Texas	Cypress Water Treatment Plant (0.33 m³/s [7.5 mgd])	2013 to 2015 (emergency water supply)	Influent to conventional water treatment facility (DPR)	TCEQ
Big Spring, Texas	Raw water production facility (0.11 m³/s [2.5 mgd])	2013	Raw water transmission main (DPR)	TCEQ
Abilene, Texas	Hamby Water Reclamation Facility (0.53 m³/s [12 mgd])	2015	Surface water reservoir (IPR)	TCEQ
City of San Diego, California	Pure Water (1.31 m³/s [30 mgd] Phase I; 2.32 m³/s [53 mgd] Phase 2)	Planning phase	Surface water reservoir (IPR)	California SWRCB

[a]ADEQ = Arizona Department of Environmental Quality; GA DNR EPD = Georgia Department of Natural Resources Environmental Protection Division; SWRCB = State Water Resources Control Board; and TCEQ = Texas Commission on Environmental Quality.
[b]Listed flow is for raw water flow to advanced WRRF unless otherwise noted.
[c]Listed capacity is for the new Goreangab Water Reclamation Facility brought online in 2001.
[d]Characterization as IPR or DPR listed in parentheses, recognizing that state and federal definitions of where projects fit along the potable reuse continuum can vary.

presence of an environmental buffer. This approach reflects recent legislation in California, which seeks to clarify the interpretation of IPR vs DPR, focusing on the continuum of potable reuse applications and the pathogen inactivation or removal required depending on the percentage of blending and retention time (California Legislature Assembly Bill 574, 2017).

7.1   Current Regulatory Framework

No federal regulations have been promulgated in the United States specifically governing water reuse; however, any water produced for potable consumption must meet all primary and secondary drinking water standards promulgated under the Safe Drinking Water Act before distribution for public consumption. A list of all NPDWRs can be found on U.S. EPA's website (https://www.epa.gov/sites/production/files/2015-11/documents/howeparegulates_mcl_0.pdf). For all states with enforcement primacy, state standards may be established that are at least as stringent as the NPDWRs and would also need to be met.

Nine states have established regulations or guidelines for IPR (U.S. EPA, 2012). California adopted regulations for groundwater replenishment IPR in 2014. Surface source augmentation IPR regulations have been drafted, and California is investigating the criteria that would be necessary for safe DPR (SWRCB, 2016). The criteria and anticipated criteria for the various potable reuse options address log-reduction requirements for specified pathogens, source control for toxic chemicals, compliance with state and federal drinking water standards, treatment to control unregulated chemicals of emerging concern, multibarrier treatment, water quality monitoring, and facility operation.

In Texas, no state regulations have been developed specific to potable reuse, but the state has reviewed and authorized operation of two DPR facilities using a case-by-case approach. An NWRI independent advisory panel was convened by the New Mexico Environment Department to develop proposed operational procedures and guidelines for the first DPR facility planned within the state of New Mexico (NWRI, 2015a). The state is considering a case-by-case approval permitting process. Colorado is working with WateReuse Colorado to develop a framework for state DPR regulations.

Requirements for removal (or inactivation) of microbial contaminants have received considerable deliberation within the states that have either promulgated regulations for potable reuse or developed requirements for specific reuse facilities. California and Texas's approach provide a good framework for how the regulatory community is approaching microbial removal requirements. Both states base their microbial removal requirements on the U.S. EPA end goal of 1 in 10,000 annual risk of infection per person or lower from consumption of drinking water derived from the Surface Water Treatment Rule. Table 4.7 lists the corresponding finished water concentrations of viruses, *Cryptosporidium,* and *Giardia.* Notably, the U.S. EPA finished water goals for pathogens are similar to the corresponding WHO drinking water guidelines based on disability adjusted life years (DALY) (WHO, 2006), illustrating consistency in an ultimate goal of minimizing human infection and potential illness from exposure to pathogens in drinking water.

The California groundwater replenishment rules require that treatment is provided to achieve 12-, 10-, and 10-log inactivation/removal of viruses, *Giardia,* and *Cryptosporidium,* respectively, based on maximum reported concentrations of corresponding pathogens in raw wastewater. Texas has established individualized pathogen removal requirements for DPR projects that are based on the same finished water goals, but depend on the measured or expected concentrations of pathogens in the secondary or final WRRF effluent used as the source to the DPR treatment facility. Where pathogen control requirements are not specified in regulation, pathogenic organisms should be controlled to achieve the tolerable risk for the jurisdiction. The densities of those organisms in wastewater have been published (e.g., Asano et al., 2007) or can be determined for the project wastewater.

To further control risks associated with exposure to microbial pathogens, the California Groundwater Replenishment Rule (GWRR) requires that any individual treatment process can only be credited with up to 6-log removal, setting the precedence for requiring *multiple* barriers for pathogens. The use of multiple and redundant barriers for pathogens has become the industry standard as reflected in language included in *Framework for Direct Potable Reuse* (NWRI, 2015b).

Chemical constituents in potable reuse projects approved to date, regardless of the state of implementation, have been regulated based on compliance with NPDWRs, any state-specific requirements (e.g., California notification levels), and guidance to provide diverse barriers to address unregulated CECs. For IPR, the treatment and water quality requirements include both

TABLE 4.7 Concentrations of pathogens in finished water corresponding to a 1 in 10,000 annual risk of infection (Trussell et al., 2013).

Pathogen	Concentration	Reference
Virus	2.2×10^{-7}/L	Regli et al., 1991
Giardia	6.8×10^{-6} cysts/L	Regli et al., 1991
Cryptosporidium	3×10^{-5} oocysts/L	Haas et al., 1996

meeting water quality objectives for potable reuse following storage in an environmental buffer, as well as meeting any requirements to prevent degradation of the quality of the receiving water. The California GWRR established criteria for total organic carbon (maximum of 0.25 mg/L in 95% of samples within the first 20 weeks), 1,4-dioxane (0.5-log reduction through an advanced oxidation process), and total nitrogen (10 mg/L as nitrogen).

The project proponent must be prepared to demonstrate to the state water regulators how the risks posed by their potable reuse project will be managed to meet the state's public health goals and comply with all state and federal drinking water standards. *Potential for Expanding the Nation's Water Supply Through Reuse of Municipal Wastewater* (NRC, 2012) and *Framework for Potable Reuse* (NWRI, 2015b) are some of the resources that can help a utility assess the risks and develop a risk management approach.

7.2   Other Legal Considerations

Central to any IPR or DPR project is planning, design, construction, and startup of an advanced treatment facility to address microbial removal requirements and standards and goals for chemical microconstituents. As with implementation of any new water treatment facility, multiple regulatory approval steps must be addressed for authorization to construct and operate a new facility. State regulations should be reviewed and meetings scheduled with the regulatory enforcement agency (Figure 4.1) to discuss specific requirements for approval of the new treatment facility. The following milestones may be required depending on state-specific regulations:

- Source water characterization to meet Long Term 2 Enhanced Surface Water Treatment Rule requirements as well as any specific requirements specific to potable reuse;
- Pilot testing technologies proposed for inclusion in the advanced treatment train;
- Plan and specification review for approval to design and construct the treatment facility;
- Submittal of a disinfectant concentration-time study for inclusion in the surface water monthly operating report if chemical disinfection credits are claimed as part of the advanced treatment facility;
- Ultraviolet validation report if UV disinfection or advanced oxidation is included in the advanced treatment facility with claims for pathogen inactivation credit;
- Full-scale verification test protocols and reports;
- Development and submittal of CCPs following the HACCP approach and/or alarm and shutdown triggers;

- Monitoring plans, including what samples are to be collected when and analyzed by what method; and
- Development and submittal of standard operating procedures and training documents.

Permitting requirements for disposal of residuals generated from the advanced WRRF must also be met. The industrial pretreatment program may be reviewed with new requirements imposed.

For IPR projects, regulations affiliated with discharge of the recycled or advanced treatment product water must be met. The UIC program requirements may apply for discharge of recycled water to an aquifer. Applicable water rights and contractual agreements affected by the diversion and potential downstream extraction of the recycled water will need to be reviewed with any relevant legal requirements addressed, as outlined in Section 2.

7.3   Risk Management and Communication Strategies

The most critical risks associated with potable reuse projects include the following:

- Public health risk associated with consumption of water with compromised microbial or chemical quality and
- Public concerns derailing a project.

The first risk is discussed herein. Chapter 6 presents guidelines on developing a public outreach and communication plan to mitigate adverse public perceptions.

The potable reuse community has invested considerable attention on the question of reliability to minimize public health risks associated with production of off-specification water, particularly in the absence of an environmental buffer for DPR projects. For microbial risks, the use of qualitative microbial risk assessment (QMRA) is gaining attention to carefully consider potential variability in pathogen loading and treatment achieved at the advanced treatment facility. Qualitative microbial risk assessment, as applied to potable reuse, uses Monte Carlo analysis to evaluate the annual risk of infection from exposure to pathogens in recycled water, accounting for (1) site-specific or reference data on the range of pathogen concentrations measured in raw or treated wastewater and (2) the range in pathogen removal achieved through unit processes included in the recycled water treatment train. The California State Water Resources Control Board (SWRCB) is recommending the use of QMRA as a requirement for DPR projects (SWRCB, 2016). The WHO is also considering inclusion of the use of QMRA for statistical analysis of the continued ability to meet DALY goals in potable reuse projects. Additional recommendations for reliability and mitigating risks from exposure to pathogens or chemical microconstituents include the following:

- Development of a source control program that includes reviewing the characteristics of industrial discharges permitted through the Industrial Pretreatment Program;
- Inclusion of multiple, independent barriers for pathogens;
- Inclusion of diverse barriers for pathogens and chemical microconstituents;
- Defined critical control points to assess process performance and operations at each step, including alarms and control for detection and response for off-specification water; and
- Adequate storage/travel time to enable necessary response based on the failure response time (Salveson et al., 2016).

The SWRCB (2016) and its expert panel also highlighted the need for non-treatment barriers to mitigate risk, including the following:

- Water resource recovery facility operation, recognizing the importance of wastewater treatment processes as the next step after sewer-shed control in controlling operations and quality;
- Demonstration of technical, management, and financial capacity; and
- Operator training.

The chemical and microbial stability of the finished water must also be reviewed, with appropriate post-treatment incorporated to minimize degradation of water quality in the distribution system and meet requirements set forth under the Lead and Copper Rule, Total Coliform Rule, and Stage 1 and 2 Disinfectants/Disinfection By-Product Rules (U.S. EPA, 1998; U.S. EPA, 2006).

A few risks associated with IPR also warrant mention. Indirect potable reuse presents the possibility that the water source being augmented (aquifer or surface source) could be contaminated because of a wastewater treatment failure. There is also the possibility that a drinking water standard could change or be added that causes previously compliant recycled water to become a degradation of the drinking water source. These situations can be a particular concern for groundwater recharge IPR, where the aquifer could be affected for decades. This risk can be mitigated by tracking the development of drinking water standards, fully characterizing the chemical quality of the source wastewater and industrial discharges, developing and operating a robust monitoring and control plan for targeted constituents, and implementing a source control program focused on chemicals of concern. For groundwater injection, the affect of oxidation–reduction potential and dissolved oxygen should also be reviewed with a hydrogeologist to prevent mobilization of any minerals (e.g., arsenic) that would require removal before reuse.

8.0 SUMMARY

The WHO has developed guidelines for water reuse and several countries have established regulations, standards, or guidelines for reuse activities within their nations. In the United States, no federal regulations for water reuse have been established; however, the U.S. EPA has developed and updated guidelines for both potable and nonpotable reuse activities. Some states have developed regulations, standards, and/or guidelines for agricultural reuse, urban reuse, environmental reuse, and IPR. U.S. EPA (2012) guidelines provide an overview of the state regulations, depending on reuse activities. *Water Reuse Issues, Technologies, and Applications* (Asano et al., 2007) also provides a good initial reference for regulatory and risk considerations associated with both potable and nonpotable reuse applications. For nonpotable reuse in buildings, WE&RF (2017) provides a risk-based framework that can be helpful for implementing decentralized reuse systems that are protective of public health. For DPR, NWRI (2015b) provides a framework that can be particularly valuable for DPR applications in states that have not yet established a protocol for approving projects for construction, startup, and operation.

In this chapter, we provide a generalized approach for identifying and addressing regulatory requirements and managing risk (Figure 4.1). Planners and managers are advised to first define the type or reuse application under consideration, review relevant regulations in their jurisdiction, and review water rights and contractual agreements affecting the quantity of recycled water available and the optimal approach for implementing the project. Water quality should be reviewed and compared to any relevant state standards and use-specific water quality requirements to manage risk. Completing these steps, planners and managers are advised to review Chapter 6 on public outreach and communications plans, which should be initiated early in a project to control messaging and perceptions. Refer to Chapter 7 to identify treatment required to address regulatory requirements and goals and refer to Chapter 8 to develop a monitoring and control plan corresponding to the HACCP approach and to mitigate risks.

9.0   REFERENCES

Alan Plummer Associates, Inc. (2015) *Final Report: Direct Potable Reuse Resource Document*; Report prepared for the Texas Water Development Board, TWDB Contract No. 1248321508.

Amoah, P.; Keraita, B.; Akple, M.; Drechsel, P.; Abaidoo, R. C.; Konradsen, F. (2011) *Low-Cost Options for Reducing Consumer Health Risks from Farm to Fork where Crops Are Irrigated with Polluted Water in West Africa*; International Water Management Institute: Colombo, Sri Lanka.

Asano, T.; Burton, F. L.; Leverenz, H. L.; Tsuchihashi, R.; Tchobanoglous, G. (2007) *Water Reuse Issues, Technologies, and Applications*; McGraw-Hill: New York.

Baronti, C.; Curini, R.; D'Ascenzo, G.; DiCoricia, A.; Gentili, A.; Samper, R. (2000) Monitoring Natural and Synthetic Estrogens at Activated Sludge Sewage Treatment Plants and in a Receiving River Water. *Environ. Sci. Technol.*, **34** (24), 5059–5066.

Benami, M.; Gross, A.; Herzberg, M.; Orlofsky, E.; Vonshak, A.; Gillor, O. (2013) Assessment of Pathogenic Bacteria in Treated Graywater and Irrigated Soils. *Sci. Total Environ.*, **458–460** (August), 298–302.

Blanky, M.; Rodríguez-Martínez, S.; Halpern, M.; Friedler, E. (2015) Legionella Pneumophila: From Potable Water to Treated Greywater; Quantification and Removal during Treatment. *Sci. Total Environ.*, **533** (November), 557–565.

California Code of Regulations (2015) Water Recycling Criteria, Title 22, Division 4, Chapter 3, California Code of Regulations: Sacramento, California.

California Legislature (2017) Assembly Bill 574. https://leginfo.legislature.ca.gov/faces/billTextClient.xhtml?bill_id=201720180AB574 (accessed June 2017).

Chen, Z.; Wu, Q.; Wu, G.; Hu, H. (2016) Centralized Water Reuse System with Multiple Applications in Urban Areas: Lessons from China's Experience. *Resour. Conserv. Recy.*, **117** (2017), 125–136.

Crook, J. (2004) *Innovative Applications in Water Reuse*. WateReuse Research Foundation. https://watereuse.org/wp-content/uploads/2015/10/WRA-101.pdf (accessed May 2017).

Engineering-Science (1987) *Monterey Wastewater Reclamation Study for Agriculture;* Final Report prepared for the Monterey Regional Water Pollution Control Agency, California.

Falas, P.; Wick, A.; Castronovo, S.; Habermacher, J.; Ternes, T. A.; Joss, A. (2016) Tracing the Limits of Organic Micropollutant Removal in Biological Wastewater Treatment. *Water Res.*, **95**, 240–249.

Florida Department of Environmental Protection (2014) Reuse of Recycled Water and Land Application, Chapter 62-610, Florida Administrative Code; Florida Department of Environmental Protection: Tallahassee, Florida.

Garcia-Cuerva, L.; Berglund, E. Z.; Binder, A. R. (2016) Public Perceptions of Water Shortages, Conservation Behaviors, and Support for Water Reuse in the U.S. *Resour. Conserv. Recy.*, **113** (October), 106–115.

Haas, C. N.; Crocket, C. S.; Rose, J. B.; Gerba, C. P.; Fazil, A. M. (1996) Assessing the Risk Posed by Oocysts in Drinking Water. *J. Am. Water Works Assoc.*, **88** (9), 131–136.

Halliwell, D.; Burris, D.; Deere, D.; Leslie, G.; Rose, J.; Blackbeard, J. (2014) *Utilization of Hazard Analysis and Critical Control Points Approach for Evaluating Integrity of Treatment Barriers for Reuse;* Final Report for WateReuse Research Foundation Project No. 03-09; WateReuse Research Foundation: Alexandria, Virginia.

Harivandi, A. (2011) Purple Gold—A Contemporary View of Recycled Water Irrigation; *U.S. Golf Assoc. Green Sect. Rec.*, 49 (45), 1–10: W.

National Research Council (1996) Use of Recycled Water and Sludge in Food Crop Production; Committee on the Use of Treated Municipal Wastewater Effluents and Sludge in the Production of Crops for Human Consumption; ISBN: 0-309-5681 1-0; National Research Council: Washington D.C.

National Research Council (2012) *Water Reuse—Potential for Expanding the Nation's Water Supply Through Reuse of Municipal Wastewater*; The National Academies Press: Washington, D.C.

National Water Research Institute (2012) Review of California's Water Recycling Criteria for Agricultural Irrigation; National Water Research Institute: Fountain Valley, California.

National Water Research Institute (2015a) *Developing Proposed Direct Potable Reuse Operational Procedures and Guidelines for Cloudcroft, New Mexico;* Independent Advisory Panel Report prepared for New Mexico Environmental Department.

National Water Research Institute (2015b) *Framework for Direct Potable Reuse*; WateReuse Research Foundation: Alexandria, Virginia.

Oregon Water Resources Department. *Use of Recycled Water.* https://www.oregon.gov/owrd/pages/mgmt_recycled.aspx (accessed Feb 2017).

Pinto, U.; Maheshwari, B. L.; Grewal, H. S. (2010) Effects of Greywater Irrigation on Plant Growth, Water Use and Soil Properties. *Resour. Conserv. Recy.*, 54 (7), 429–435.

Prosser, R. S.; Sibley, P. K. (2015) Human Health Risk Assessment of Pharmaceuticals and Personal Care Products in Plant Tissue Due to Biosolids and Manure Amendments, and Wastewater Irrigation. *Environ. Int., 75,* 223–233.

Regli, S., Rose, J. B.; Haas, C. N.; Gerba C. P. (1991) Modeling the Risk from Giardia and Viruses in Drinking Water. *J. Am. Water Works Assoc.,* **83** (11), 76–84.

Rochelle, M. (2008) *Wastewater Wars: Implications of Reuse Projects in Texas;* White Paper, July 10, 2008.

Salveson, A.; Rauch-Williams, T.; Dickenson, E.; Drewes, J.; Drury, D.; McAvoy, D.; Snyder, S. (2012) *A Proposed Suite of Indicators for Assessing the Efficiency of Secondary Treatment for the Removal of Organic Trace Compounds;* Water Environment Research Foundation Project CEC4R08 Final Report; Water Environment Research Foundation: Alexandria, Virginia.

Salveson, A.; Steinle-Darling, E.; Trussell, R. S.; Pecson, B.; Macpherson, L. (2016) *Guidelines for Engineered Storage for Direct Potable Reuse;* Final Report for WateReuse Research Foundation Project 12-06; WateReuse Research Foundation: Alexandria, Virginia.

Scott, C.; Drechsel, P.; Raschid-Sally, L.; Bahri, A.; Mara, D.; Redwood, M.; Jimenez, B. (2010) Wastewater Irrigation and Health: Challenges and Outlook for Mitigating Risks in Low Income Countries. In *Wastewater Irrigation and Health: Assessing and Mitigating Risks in Low Income Countries;* Drechsel, P., Scott, C. A., Raschid-Sally, L., Redwood, M., Bahri, A., Eds.; Earthscan: London, U.K.

Sedlak, D.; Kavanaugh, M. (2006) *Removal and Destruction of NDMA and NDMA Precursors during Wastewater Treatment;* Final Report for WateReuse Research Foundation Project 01-002; WateReuse Research Foundation: Alexandria, Virginia.

Southwest Florida Water Management District (2009) *Recycled Water: A Reliable, Safe Alternative Water Supply;* Southwest Florida Water Management District: Brooksville, Florida.

State Water Resources Control Board (2016) *Evaluation of the Feasibility of Developing Uniform Water Recycling Criteria for Direct Potable Reuse;* State Water Resources Control Board: Sacramento, California.

Trussell, R. R.; Salveson, A.; Snyder, S. A.; Trussell, R. S.; Gerrity, D.; Pecson, B. M. (2013) *Potable Reuse: State of the Science Report and Equivalency Criteria for Treatment Trains;* Final Report for WateReuse Research Foundation Project 11-02-2; WateReuse Research Foundation: Alexandria, Virginia.

UN Water (2014) Water for Food. Update October 7, 2014. http://www .unwater.org/statistics/statistics-detail/en/c/246663/ (accessed Nov 2016).

U.S. Department of Agriculture (2012) Guide for Organic Crop Producers. www.ams.usda.gov/nop (accessed Nov 2016).

U.S. Environmental Protection Agency (1998) Stage 1 Disinfectants and Disinfection Byproducts Rule. *Fed. Regist.*, 63:241:69390.

U.S. Environmental Protection Agency (2003) *Cross-Connection Control Manual;* EPA-816/R-03-002; U.S. Environmental Protection Agency: Washington, D.C.

U.S. Environmental Protection Agency (2006) Stage 2 Disinfectants and Disinfection Byproducts Rule. *Fed. Regist.*, 71:3:388.

U.S. Environmental Protection Agency (2012) *Guidelines for Water Reuse;* EPA-600/R-12-618; U.S. Environmental Protection Agency, Office of Wastewater Management: Washington, D.C. http://nepis.epa.gov/Adobe/ PDF/P100FS7K.pdf (accessed Feb 2017).

Virginia Department of Environmental Quality (2014) Water Reclamation and Reuse Regulation (9VAC25-740); Virginia Administrative Code, Title 9, Chapter 740; Virginia Department of Environmental Quality: Richmond, Virginia.

Washington State Department of Ecology (1997) *Water Reclamation and Reuse Standards;* Washington State Department of Ecology: Olympia, Washington.

Water Environment & Reuse Foundation (2017) *Risk-Based Framework for the Development of Public Health Guidance for Decentralized Non-Potable Water Systems;* SIWM10C15; Water Environment & Reuse Foundation: Alexandria, Virginia.

Water Research Foundation (2014) *Blueprint for Onsite Water Systems: A Step by Step Guide for Developing a Local Program to Manage Onsite Water Systems,* Project No. 4580: Water Research Foundation: Denver, Colorado.

WateReuse Association; Water Environment & Reuse Foundation (2017) Personal communication; WERF & WateReuse List of Potable Reuse in U.S., Microsoft Excel database provided by Deirdre Finn; March 31.

World Health Organization (2006) *Guidelines for the Safe Use of Wastewater, Excreta and Greywater,* 3rd ed.; World Health Organization: Geneva, Switzerland.

Wu, X.; Conkle, J. L.; Ernst, F.; Gan, J. (2014) Treated Wastewater Irrigation: Uptake of Pharmaceutical and Personal Care Products by Common Vegetables under Field Conditions. *Environ. Sci. Technol.*, 48 (19), 11286–11293.

Wu, X.; Dodgen, L. K.; Conkle, J. L.; Gan, J. (2015) Plant Uptake of Pharmaceutical and Personal Care Products from Recycled Water and Biosolids: A Review. *Sci. Total Environ.*, **536**, 655–666.

Yamamoto, G. H. (2010) California Department of Public Health letter to Assistant Executive Officer K.D. Landau, Central Valley Water Board, June 15.

Yu, Z. L.; Rahardianto, A.; DeShazo, J. R; Stenstrom, M. K.; Cohen, Y. (2013) Critical Review: Regulatory Incentives and Impediments for Onsite Graywater Reuse in the United States. *Water Environ. Res.*, **85**, 650–662.

10.0 SUGGESTED READINGS

National Research Council (1998) *Issues in Potable Reuse: The Viability of Augmenting Drinking Water Supplies with Recycled Water*; The National Academies Press: Washington, D.C.

National Resource Management Ministerial Council; Environment Protection and Heritage Council; Australian Health Ministers' Conference (2009) *National Water Quality Management Strategy Australian Guidelines for Water Recycling: Managing Health and Environmental Risks (Phase 2): Managed Aquifer Recharge*; Canberra, Australia. http://www .environment.gov.au/water/quality/publications/nwqms-australian-guidelines-water-recycling-managing-health-phase2-aquifer-recharge.

U.S. Environmental Protection Agency (2006) Long Term 2 Enhanced Surface Water Treatment Rule. *Fed. Reg.*, 71:3:653.

U.S. Environmental Protection Agency (2017) National Primary Drinking Water Regulations. https://www.epa.gov/sites/production/files/2015-11/documents/howeparegulates_mcl_0.pdf (accessed Feb 2017).

5

Financial Sustainability

Thierry A. Boveri, CGFM; Robert J. Ori, and Nicholas T. Smith

Investment in water reuse projects is being driven by the need for water supply augmentation/resiliency, environmental sustainability, improved agricultural production, reduced energy consumption (i.e., water-energy nexus; for more information, please reference *The Water-Energy Nexus: Challenges and Opportunities* [Bauer et al., 2014]), and the regulation of wastewater discharges to waterbodies. Because of these motivations, water reuse is increasingly being discussed or evaluated by stakeholders in both the public and private sectors. There is significant potential for growth in water reuse considering only 20% of the total possible municipal wastewater effluent production in the United States is beneficially reclaimed (Bastian & Murray, 2012). As discussed in other chapters, water reuse options can vary from nonpotable to potable reuse and result in differing costs and benefits achieved. As a result, proper engineering, financial, and management planning is necessary to assess the most viable option. In this chapter, we seek to provide an overview of financial sustainability by defining what it is, best management practices to promote it, and the approach to using long-range strategic business planning to achieve it. Please note that this chapter focuses on the more common perspective of the utility as it relates to water reuse investment. Therefore, this chapter also discusses current issues facing the utility industry, role of effective utility management, benchmarking of reuse rate/costs, and a sample case study. In closing, this chapter seeks to address the water reuse value proposition.

1.0 FINANCIAL SUSTAINABILITY AND ISSUES FACING THE UTILITY INDUSTRY

Water and wastewater utilities are facing increasing and broadening challenges that will fundamentally affect their ability to provide for the health, safety, and welfare of the public. These challenges may be unique to a region such as prolonged droughts affecting water supplies, while others may be more universal such as the need for infrastructure reinvestment estimated to cost the United States a trillion dollars for water line replacements alone over the next 25 years (AWWA, 2012). While assistance from federal and state grant and loan programs can help mitigate the initial costs associated with required infrastructure to address these challenges, ultimately the costs will result in increasing user fees. The U.S. Bureau of Labor Statistics reports that the water and sewer maintenance index, a proxy for the increase in water and wastewater bills nationally, has averaged approximately 5% per year for the past 20 years. This rate of increase in water and wastewater bills is more than double the rate of inflation and is raising awareness of affordability concerns. To address these challenges and ensure the financial sustainability of operations, utility managers must consider a wide range of options to find the best solution. As addressed in other chapters of this roadmap, implementation of a water reuse program may provide a solution to some of these challenges depending on the circumstances of a utility and its community.

2.0   WHAT IS FINANCIAL SUSTAINABILITY?

In 2012, the World Bank issued a report on financial sustainability for bank-financed water, wastewater, and irrigation projects in developing countries. It defined *financial sustainability* as the adequacy of revenues to cover the near-term and long-term costs of service, including both operational and capital expenses, as shown in Figure 5.1. The report highlighted a prior independent evaluation of the progress and issues associated with projects spanning a 10-year period from 1997 through 2007. The evaluation noted that *"... the borrowing countries have not yet sufficiently tackled several tough but vital issues, among them broadening access to sanitation, fighting pollution, restoring degraded aquatic environments, monitoring and data collection and cost recovery"*. A significant challenge cited in addressing these issues is the negative feedback cycle that results from inadequate cost recovery. Poor cost recovery over time will lead to failing infrastructure, higher costs, and degraded service. The inefficient and higher cost of operation increases pressure on affordability and lack of willingness to pay for service, which, in turn, continues the cycle (McPhail et al., 2012).

FIGURE 5.1 Adequacy of cost recovery.

Although adequate cost recovery is an important element to financial sustainability, other considerations should be made to the effective management of operations. The framework of effective utility management (EUM), developed by six industry associations in collaboration with the U.S. Environmental Protection Agency (U.S. EPA), identifies 10 attributes that support the concept of financial sustainability; these are listed in Table 5.1 (APWA et al., 2017).

TABLE 5.1   Ten attributes of EUM (http://www.watereum.org).

1. Product Quality	6. Infrastructure Stability
2. Customer Satisfaction	7. Operational Resiliency
3. Employee Leadership Development	8. Community Sustainability
4. Operational Optimization	9. Water Resource Adequacy
5. Financial Viability	10. Stakeholder Understanding and Support

The EUM attributes related to financial viability include the following:

- Understanding and planning for life cycle costs;
- Addressing and monetizing the identified benefits of the project, including avoided costs of alternatives;
- Balancing long-term debt, asset values, operations and maintenance expenditures, and operating revenues;
- Establishing predictable and adequate rates to support current needs and future needs and taking into account affordability and the needs of disadvantaged households; and
- Identifying opportunities for diversifying revenue and raising capital.

Application of the financial viability attributes and EUM framework is further discussed in the following section of this chapter.

2.1   Developing the Strategic Business Plan

Internal competition for resources related to operating or capital funding and the proper prioritization of capital improvements requires that utility managers make informed decisions about identifying the mission's critical, most beneficial, and cost-effective capital investments. The development of the strategic business plan (SBP) is considered an essential requirement to management success pursuant to the EUM framework because of the strategic nature and importance of infrastructure investment to a utility. The SBP is a comprehensive planning exercise and decision support tool that incorporates the strategic plan, regulations, and risk assessment, as previously discussed in Chapters 3 and 4, as well as asset management and long-range financial planning. Pursuant to the EUM framework, the SBP should do the following:

- Assess current conditions by conducting a strengths, weaknesses, opportunities, and threats analysis;
- Characterize a continuum of possible and likely future conditions (i.e., sensitivity analysis);
- Assess underlying causes and effects of future conditions; and
- Establish a vision, objectives, strategies, and underlying organizational values.

Ultimately, the SBP should be flexible and incorporate any number of decision support tools as previously noted. The following provides a further discussion of best practices for asset management programs and long-range financial planning.

2.1.1 Asset Management Program and Existing Infrastructure Priorities

Before endeavoring to invest in new infrastructure, a utility is charged with maintaining its existing infrastructure. Pursuant to the U.S. EPA best practices guide, *"Asset management is maintaining a desired level of service for what you want your assets to provide at the lowest life cycle cost. Lowest life cycle cost refers to the best appropriate cost for rehabilitating, repairing, or replacing an asset. Asset management is implemented through an asset management program and typically includes a written asset management plan"* (U.S. EPA, 2008). Pursuant to the U.S. EPA asset management framework, a utility should seek to develop an asset management program based on the five core questions and best practices, as illustrated in Table 5.2.

The results and findings of the asset management program are valuable because they can aid a utility in identifying the best and most efficient means of asset management and replacement.

2.1.2 Long-Range Financial Planning

Developing a long-range financial plan will assist the utility in addressing several key elements of financial sustainability including identification of life cycle costs, costs to the environment and community, testing sensitivity of changes in customer base or operations, developing a plan of finance to fund the identified capital needs, determination of projected compliance with financial policies and loan or bond covenants, and identification of associated user fee rate adjustments and customer effects. The planning horizon should be established based on the life cycle of the underlying asset to ensure full accounting of periodic and replacement costs.

The long-range planning model is a holistic evaluation of business operations to determine how incremental effects from new programs, for example, may affect the fiscal position of the utility. The following methodology for assessing revenue sufficiency is a commonly accepted practice and is identified in greater detail within the Water Environment Federation's (WEF's) Manual of Practice No. 27, *Financing and Charges for Wastewater Systems* (2004). Figure 5.2 provides an illustration of the methodology to assess the revenue sufficiency of utility operations.

As Figure 5.2 shows, the source data and general user inputs then flow to the customer statistics, which are then applied to the rates for service to calculate revenues under existing rates. Other operating revenues and available cash reserves are then included together with the rate revenues under existing rates to identify the total available resources to fund the revenue requirements of the utility. Revenue requirements can be defined as the cost of operations, debt service, and other funding requirements less any income

TABLE 5.2 U.S. EPA five core questions and best practices for asset management (U.S. EPA, 2008).

The five core questions to asset management	Best practices
1. Assessing the current state of assets?	• Preparing an asset inventory and system map • Developing a condition assessment and rating system • Assessing remaining useful life by consulting projected-useful-life tables or decay curves • Determining asset values and replacement costs
2. What is my required "sustainable" level of service?	• Analyzing current and anticipated customer demand and satisfaction with the system • Understanding current and anticipated regulatory requirements • Writing and communicating to the public a level-of-service "agreement" that describes your system's performance targets • Using level-of-service standards to track system performance over time
3. What assets are critical to sustained performance?	• Listing assets according to how critical they are to system operations • Conducting a failure analysis (root cause analysis, failure mode analysis) • Determining the probability of failure and listing assets by failure type • Analyzing failure risk and consequences • Using asset decay curves • Reviewing and updating your system's vulnerability assessment (if your system has one)
4. What are my minimum life cycle costs?	• Moving from reactive maintenance to predictive maintenance • Knowing the costs and benefits of rehabilitation vs replacement • Looking at life cycle costs, especially for critical assets • Deploying resources based on asset conditions • Analyzing the causes of asset failure to develop specific response plans
5. What is my best long-term funding strategy?	• Revising the rate structure • Funding a dedicated reserve from current revenues (i.e., creating an asset annuity) • Financing asset rehabilitation, repair, and replacement through borrowing or other financial assistance

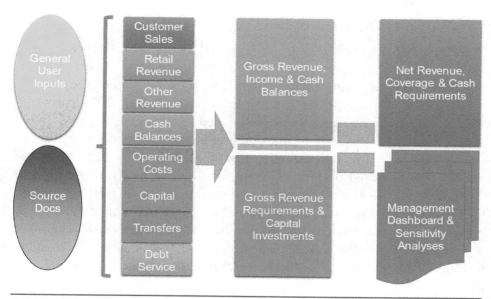

FIGURE 5.2   Long-range planning model flow chart and methodology.

and funds from other sources that, in turn, result in the net revenue requirements to be funded from rate revenues. A deficiency to net revenues would indicate the need for a rate adjustment, but can also indicate a financial sustainability or viability issue. The following list is intended to provide utility managers with a better understanding of the necessary data required to develop the long-range financial plan:

- Historical customer billing statistics and adopted user rate schedules to model user fee revenues;
- Cash balances for the most recently completed fiscal year to determine funds available to fund capital needs and maintain operating reserves;
- Current operating budget, which would provide a baseline for existing operating expenses and budgeted transfers to general fund or other revenue requirements;
- Debt service schedules for all outstanding debt;
- Capital improvement projects and estimated capital costs, including any estimates for capital needs identified by the asset management program and any estimates of incremental operating expenses associated with changes to utility infrastructure; and
- Other miscellaneous data that may be relied upon such as fixed asset records for the allocation of certain costs.

2.1.2.1   Water Reuse Pricing and Utility Revenues

The majority of the U.S. population is served by public utilities, which are typically established as enterprise funds whereby user fees are typically the primary source of revenues. Approximately 84% of the U.S. population receives water through public water systems (Maupin et al., 2010), of which 89% is served by public/government-owned facilities (Beider and Tawil, 2002). Similarly, over 75% of the U.S. population is served by centralized wastewater collection and resource recovery facilities (U.S. EPA, 2004), of which 97% is served by public/government-owned facilities (Beider and Tawil, 2002).Typical fees charged by utilities include user fees, which may include a fixed fee per account and variable charge per unit of billed flow; growth-related fees such as system development charges, connection fees, and tap fees; and other miscellaneous fees or charges for service. In some instances, utilities may assess the property for the extension of service. Specifically with respect to water reuse pricing for retail service, user fees are often similar to that of retail water and wastewater fees; however, water reuse user fees may or may not recover the full cost of service. Water reuse pricing is affected by three primary factors: benefits, costs, and politics. The type of water reuse, potential customers, and identified benefits will affect the pricing strategy. For direct potable water reuse, the benefit is no different from other sources of potable water, with the exception of the likely differences in the costs that may affect pricing. By contrast, nonpotable water reuse irrigation may provide a distributed benefit among both the water and wastewater systems, and, recognizing that it often takes effluent from multiple residential households to provide enough irrigation demand for one residential home, not all customers are likely to have access to the service. Rate setting is a political process, which is not predictable and can be affected by economic cycles and local issues. For more information about cost of service, rate structures, and pricing, refer to WEF's Manual of Practice No. 27, *Financing and Charges for Wastewater Systems* (2004).

2.1.2.2   Grant Funding and Low Interest Loans

The federal government, states, and local agencies work together to offer utilities assistance with financing for capital projects such as water reuse programs. Grant funding is unique to each state and requires that a utility contact their appropriate state environmental regulatory agency. Low interest rate loans are through a state revolving loan program, or pooled bond financing may be available. Table 5.3 provides a listing of federally funded loan programs, which may be administered through the state environmental regulatory agency.

It is recommended that utilities considering the issuance of additional indebtedness discuss financing options with their financial advisor.

TABLE 5.3 Low interest loan programs and eligibility.

Federal agency	Program	Eligible projects
U.S. EPA	Drinking Water State Revolving Fund Program	Construction of facilities that will facilitate compliance with national primary drinking water regulations or further the health protection objectives of the Safe Drinking Water Act.
U.S. EPA	Clean Water State Revolving Fund Program	Eligible projects may include a wide range of water quality projects, such as water resource recovery facilities (WRRFs); agricultural, rural, and urban runoff control estuary improvement projects; and wet weather flow control groundwater protection projects.
U.S. EPA	Water Infrastructure Finance and Innovation Act (WIFIA)	• $20 million: minimum project size for large communities. $5 million: minimum project size for small communities' population of 25,000 or less). 49%: maximum portion of eligible project costs that WIFIA can fund. Total federal assistance may not exceed 80% of a project's eligible costs. • 35 years: maximum final maturity date from substantial completion. 5 years: maximum time that repayment may be deferred after substantial completion of the project. Interest rate will be equal to or greater than the U.S. Treasury rate of a similar maturity at the date of closing. Projects must be creditworthy and have a dedicated source of revenue. • NEPA, Davis-Bacon, American Iron and Steel, and all other federal cross-cutter provisions apply.
Bureau of Reclamation U.S. Department of Interior (DOI)	Water Conservation Field Services Program/Efficiency Incentives Program	Eligible recipients generally include water systems that contract for water supplies through the Bureau of Reclamation.
Bureau of Reclamation DOI	WaterSMART	Funding for water reuse engineering and planning studies may be available to your utility through federal or state agencies such as the U.S. DOI's WaterSMART initiative.
Rural Utilities U.S. Department of Agriculture	Pop. <10,000	The program primarily funds the construction of drinking water and wastewater infrastructure (approximately 60% of assistance is allocated to drinking water improvements).

2.1.2.3   Capital Improvement Plan and Other Finance Options

The capital improvement plan (CIP) should include all identified capital needs and reflect both the replacement of existing infrastructure associated with the asset management program, expansion-related improvements based on assumed growth in the projected customer base, any applicable regulatory improvements, and any proposed improvements for sustainability (i.e., reuse water program) and efficiency. Projects should be identified by utility system (i.e., water/wastewater) and should give consideration to adjustment for inflation. During development of the plan of finance for the CIP, the funding sources can be characterized under any of the primary classifications, as follows:

- Existing cash reserves—should be used as a first priority to minimize the need for additional debt for the long-term sustainability of the utility rates;
- Expansion-related fees—should be used to pay expansion related debt as a first priority, if possible, and then expansion related capital;
- Rate revenues or "pay-go" funding—this funding source should be used for renewals, replacements, and improvements;
- Additional debt—repayment term for the loan should match the service life of the asset being financed. If possible, prioritize financing for expansion-related capital. The timing and terms of the debt should be discussed with the respective utility's financial advisor; and
- Grants or contributions from other entities or reginal partners.

2.1.2.4   Operating Expenses

The development of operating expense projections should be adjusted for inflation and changes in customer population and flow. Costs should be grouped by budget function (i.e., administration, water treatment, field operations, water reuse, etc.) and by type of expense (i.e., personnel, electric, etc.) and allocated by the respective system (i.e., water, wastewater and/or water reuse). The forecast of operating expenses should consider the need for additional personnel or other incremental operating expenses.

With respect to identifying changes in costs from the implementation of a water reuse program, there are several factors that affect costs including the following: the location of a reclaimed water source (i.e., the WRRF), treatment infrastructure, facility influent water quality, customer use requirements, transmission and pumping, timing and storage needs, energy requirements, concentrate disposal, permitting, and financing costs (NRC, 2012).

2.1.2.5 Identifying Financial Performance Measurements

Continual measurement and tracking of financial condition and performance is important to ensure that a utility has the financial resources available to meet the planned and, at times, unplanned requirements. Developing a formal or informal policy to identify minimum financial performance measures allows a utility to better compare alternatives during long-range planning evaluations. The policy should identify objectives such as minimizing risk, providing adequate working capital for operations, balancing the use of debt relative to internal sources of funding, and promoting capital reinvestment for the sustainability of operations. Key metrics often relied upon by the rating agencies in assessing the creditworthiness of a utility are a good reference point.

2.1.2.6 Revenue Sufficiency, Customer Impact, and Affordability

With increasing challenges and costs, it can reasonably be expected that utilities will continue to raise rates. Assessing revenue sufficiency is a measure of the adequacy of utility rates and revenues in funding the necessary financial resources for the financial sustainability of operations. Through long-range planning, a utility can assess forecasted deficiencies to cash flows from the issuance of debt, increases in operating expenses, or other factors such as a change in customer base or sales. The financial targets provide guidance as to the adequacy of cash reserves and cash flows during long-range planning exercises. Deficiencies to projected cash flows can be addressed through identified rate adjustments. Establishing baseline financial projection will allow for comparison and evaluation of the incremental effects from the identified capital and operating costs and potentially additional revenues from a new service associated with implementation of a water reuse program.

If additional rate adjustments are identified, the effects on customers and affordability should be evaluated. Establishing a sampling of the average or median customer's bill to assess effects can be instructive as to the general effect on the majority of the customers of a utility. In assessing affordability, reference is often made to the U.S. EPA framework used for analyzing the affordability of federal mandates stemming from the Clean Water Act and Safe Drinking Water Act. The general finding is that effects greater than 2% of the median household income per an individual water or wastewater system would present an affordability concern (USCM et al., 2013). However, income does not typically follow a normal distribution curve and, therefore, median household income may not be appropriate for determining affordability for all customers of a utility. In reality, affordability is a more fluid and interdependent function of an individual's or household's income and expenses. Those that are more susceptible to concerns of affordability would be those with the lowest income per capita per household. Therefore, any affordability evaluations should be expanded to assess the distributions of household income below the median within a given community. For additional information regarding affordability-related issues, refer to *Best Practices in Customer Payment Assistance Programs* (Cromwell et al., 2010).

2.2 Reuse Cost Benchmarking Survey and Findings

Benchmarking is a valuable tool for assessing a utility's operational efficiency relative to other utilities. Although dated, the American Water Works Association published a comprehensive water reuse rate and charge survey (Carpenter et al., 2008). The survey compared the change from 2000 through 2007. The initial survey was sent to approximately 500 utilities with 109 respondents, and followed up in 2007 to 89 of the 109 original respondents with approximately 30% of those surveyed in 2007 responding. The majority of the respondents were from California and Florida, which is generally consistent with reports by the U.S. EPA and the Florida Department of Environmental Protection with respect to the states with the most reported water reuse programs. The principal use for water reuse was reported to be for golf courses with public right-of-ways, park/landscape irrigation, construction activity, or power plant/industrial cooling processes. The primary use of reclaimed water is for irrigation to offset potable water use. The principal findings of the study are presented in Table 5.4.

A notable finding from the survey was that approximately 40% of survey respondents recovered less than 25% of costs. This is most likely indicative of reuse water sold for nonpotable irrigation use, which typically has a low value.

2.3 Sample Case Study on Financial Sustainability and the Water Reuse Value Proposition—Alexandria Renew Enterprises

Quick Highlights:
- Water resource recovery facility located in Alexandria, Virginia
- Provides service to approximately 320,000 people in the City of Alexandria and Fairfax County
- Processes 35.6 mgd average daily flow of influent
- In process of Leadership in Energy and Environmental Design Platinum Certification of the Environmental Center

- Construction of soccer field above 18-mil. gal nutrient management facility

TABLE 5.4 Water reuse rates and charges survey (Carpenter et al., 2008).

Survey year	2000	2007
Respondents system description:		
Retail	41%	N/A
Wholesale	27%	N/A
Both	32%	N/A
Cost recovery of rates less than 25%	40% approximately	40% approximately
Reuse program funding sources		
Subsidies/grants	26%	18%
Supported by water rates	31%	24%
Wastewater/combined/other	43%	58%
Basis of rate design		
Promote use of water reuse	24%	42%
Based on market conditions	9%	5%
Cost of service	14%	11%
Percent of potable water	19	16
No charge	16.5%	N/A
Contractual	7.5%	N/A
Other	10%	26%

Alexandria Renew Enterprises (AlexRenew) is a regional water resource recovery authority located in Alexandria, Virginia, a suburb of Washington D.C., and was created in a partnership between the City of Alexandria and Fairfax County. In 2011, AlexRenew began an approximately $160 million capital improvement program, referred to as the *State-of-the-Art Nitrogen Upgrade Program* (SANUP), to raise the standard of wastewater treatment quality to address limit of technology requirements to meet federal and state regulations and improve the quality of the Potomac River and Chesapeake Bay. Leading up to implementation of the project, AlexRenew developed a long-range SBP to assess the financial viability and sustainability of the project. The SBP is updated regularly and incorporated financial projections of the capital and operational changes to operations, discussed in further detail here. Although AlexRenew ultimately required a multiyear rate adjustment plan to fund the improvements, the use of debt and receipt of over $30 million in grant funding helped minimize the immediate financial burden to existing customers.

The SANUP program included four packages, with two notable elements of the project being to help lower energy consumption and provide a local community benefit through the addition of a multi-purpose athletic field, as described in greater detail as follows:

- Package B—Centrate Pretreatment Facility (CPT): This facility is the first designed and separately constructed full-scale sidestream deammonification system in North America. Inside the CPT, Anammox® bacteria, also known as "red bugs", work in tandem with other types of microorganisms to transform ammonia–nitrogen in dirty water into harmless nitrogen gas. This revolutionary deammonification process uses less air, energy, and chemicals than traditional treatment processes and offers opportunities for operation and maintenance cost savings; and

- Package C—Nutrient Management Facility: This 18-mil.-gal facility helps balance the amount of nitrogen that goes into AlexRenew's biological reactor basins. On top of the facility is Limerick Street Field. This multipurpose athletic field is a community amenity that provides a green gathering place for residents. AlexRenew formally presented the field to the City of Alexandria on October 23, 2015. The City of Alexandria will maintain the field.

In conjunction with the SANUP improvements AlexRenew made improvements to its UV disinfection system, reclaimed water booster pumping stations, and distribution lines to enable AlexRenew to produce and sell reclaimed water. Reclaimed water is a component of AlexRenew's plan to reach effluent guidelines and protect the potable water supply needed today and in the future. Providing reclaimed water creates a partnership with the community, ensuring a dependable, resilient utility providing high-quality water services at an affordable price. The use of reclaimed water helps to preserve potable water supplies by providing a reliable and economical alternative source of water used by the community other than for drinking (Pallansch and Corning, 2015).

2.4   Water Reuse Value Proposition and Conclusion

The value proposition of water reuse will vary by community and by type of reuse option (Bastian and Murray, 2012). Measuring the value proposition can be determined on the basis of relative benefits to costs. The use of proper planning, financial sustainability considerations, and effective management, as discussed in this chapter, among other things, all serve to aid decision-makers in validating the water reuse value proposition. Specifically, long-range planning to assess the benefits of water supply augmentation through indirect potable reuse or direct potable reuse can represent a significant value proposition from the avoided cost of constructing a new WRRF. Moreover, and as evidenced by the previous case study, innovative or creative solutions, such as the addition of green space in an urban environment, can accrue significant and intangible benefits back to a local community. The value proposition is only expected to increase over time. For example, when a severe drought affected more than one-third of the United States in 2012, some power plants and other energy production activities were constrained by the lack of water supply (Bauer et al., 2014). It is reasonable to assume climate change, erratic weather, population migrations, increasing energy demands, and other factors can be expected to increase the value proposition of water reuse over time in the future.

3.0 REFERENCES

American Water Works Association (2012) *Buried No Longer: Confronting America's Water Infrastructure Challenge;* American Water Works Association: Denver, Colorado.

American Public Works Association; American Water Works Association; Association of Clean Water Administrators; Association of Metropolitan Water Agencies; Association of State Drinking Water Administrators; National Association of Clean Water Agencies; National Association of Water Companies; U.S. Environmental Protection Agency; Water Environment Federation; Water Environment & Reuse Foundation; Water Research Foundation (2017) *Effective Utility Management: A Primer for Water and Wastewater Utilities;* U.S. Environmental Protection Agency: Washington, D.C.

Bastian, R.; Murray, D. (2012) *2012 Guidelines for Water Reuse;* U.S. Environmental Protection Agency, Office of Research and Development: Washington, D.C.

Bauer, D.; Philbrick, M.; Vallario, B. (2014) *The Water-Energy Nexus: Challenges and Opportunities;* U.S. Department of Energy: Washington, D.C.

Beider, P.; Tawil, N. (2002) *Future Investment in Drinking Water and Wastewater Infrastructure;* Congressional Budget Office: Washington, D.C.

Carpenter, G. W.; Grinnell, G. K.; Haney, C. M.; Jacobi, G. A.; Koorn, S. W.; O'Reilly, D. M.; . . . Vandertulip, D. (2008) *Water Reuse Rates and Charges 2000 and 2007 Survey Results;* American Water Works Association: Denver, Colorado.

Cromwell, J. E.; Colton, R. D.; Rubin, S. J.; Herrick, C. N.; Mobley, J.; Reinhardt, K.; Wilson, R. (2010) *Best Practices in Customer Payment Assistance Programs;* Water Research Foundation: Washington, D.C.

Maupin, M.; Kenny, J.; Hutson, S.; Lovelace, J.; Barber, N.; Linsey, K. (2010) *Estimated Use of Water in the United States in 2010*; U.S. Geological Survey: Reston, Virginia.

McPhail, A.; Locussol, A. R.; Perry, C. (2012) *Achieving Financial Sustainability and Costs in Bank Financed Water Supply and Sanitation and Irrigation Projects*; Water Unit, Transport, Water and ICT Department, Sustainable Development Vice Presidency: Washington, D.C.

National Research Council (2012) *Water Reuse: Potential for Expanding the Nation's Water Supply Through Reuse of Muncipal Wastewater*; The National Academies Press: Washington, D.C.

Pallansch, K.; Corning, B. (2015) Engaging the Community: Alexandria Renew Enterprises Implements a Reclaimed Water Program. *Proceedings of the Water Environment Federation, Utility Management*; p 3.

U.S. Conference of Mayors; The American Water Works Association; The Water Environment Federation (2013) *Affordability Assessment Tool for Federal Water Mandates*.

U.S. Environmental Protection Agency (2004) *Primer for Muncipal Wastewater Treatment Systems*; U.S. Environmental Protection Agency, Office of Water and Office of Wastewater Management: Washington, D.C.

U.S. Environmental Protection Agency (2008) *Asset Management: A Best Practices Guide*; EPA 816-F-08-014; U.S. Environmental Protection Agency: Washington, D.C.

Water Environment Federation (2004) *Financing and Charges for Wastewater Systems*; Manual of Practice No. 27; Water Environment Federation: Alexandria, Virginia.

Communication and Outreach

Patricia Tennyson and A. Brooke Wright

1.0 INTRODUCTION

Water has been used and reused since the beginning of time, but public understanding and awareness of this fact is not high. Projects that reclaim and recycle water for irrigation and industrial use have been operating in various states for more than 100 years. Survey research shows widespread and strong support for nonpotable recycled water use for irrigation and industrial applications. Even so, certain reuse applications—agricultural crops, for example—are not accepted in some locations. And although de facto reuse occurs on every river system in the country (National Academy of Sciences, 2012), it is only relatively recently that potable reuse projects have begun to proliferate. Robust public outreach and communication programs are needed to inform community members about the potential benefits of water reuse projects and to ensure that these projects are judged on a level playing field against other water supply projects. This chapter provides a high-level introduction to communicating about water reuse. More in-depth discussions of outreach strategies and activities are available, perhaps most notably at watereuse.org.

The increasing need for reliable, sustainable, and resilient water supplies is driving all types of alternative water supply projects, including water reuse projects. Alternate water sources such as desalination and water reuse are receiving more attention than ever before. In addition, drought awareness was never higher than it was in California in 2016, and lengthy drought patterns, particularly in Texas and California, have raised drought awareness nationwide. In addition, there appears to be more public awareness regarding the fact that droughts reoccur. Although water use efficiency is widely implemented, a region cannot always conserve its way out of a drought. Because desalination is not an option for all areas, even those on coastlines, and because conservation will not solve all supply problems, more water reuse projects are being considered, planned, and constructed.

Any water project might face questions and raise concerns among community members. And the reality is that all water supply projects compete for financial resources with all of the other infrastructure and community needs. Add to that the fact that wastewater (or *sewer water* as the Orange County Water District in Fountain Valley, California, called its source water) is the source for recycled water no matter what its final use, water reuse has some unique communication and outreach challenges. How proponents of water reuse projects communicate with their customers, community, and stakeholders about water reuse is arguably *the* most critical factor in whether the project is implemented.

A successful community outreach and communication program is based on trust, transparency, and consistent/sustained outreach efforts. Community outreach and engagement programs must be adequately funded and staffed to ensure this important local water supply source will be fairly evaluated as the community chooses how to spend its scarce financial resources. The project purpose and need must be clear and easy to understand, and then careful strategic planning is needed—both from an engineering and communication standpoint—for a water reuse project to become a reality.

Potable reuse projects are becoming more common throughout the United States as the next step for water reuse. The one universal truth when it comes to potable reuse is that lack of technology to reclaim or even purify wastewater is not typically what keeps a potable reuse project from being constructed. If a potable reuse project does not move forward, it is most likely because of public or political concerns, regulatory issues, or economic feasibility. Fortunately for the utility or other entity proposing the potable or nonpotable reuse project, there are successful models to follow and step-by-step instructions on how to talk to the public about water reuse, both nonpotable and potable. Again, this chapter provides a snapshot of how to communicate about water reuse and lists some further references for those who want to know more.

2.0 COMMUNICATING ABOUT WATER REUSE

A solid public outreach program with a written strategic communication plan is the first step toward success for any infrastructure project, including water reuse projects. The outreach plan should include traditional and "new" strategies and tactics to engage community interest and provide easy-to-understand information to community members of all ages. In addition, it is important to note that the development of a purpose and need statement, as discussed in Chapter 2, and engaging in strategic planning and concept development, as described in Chapter 3, help lay the groundwork for an effective communication and stakeholder engagement effort.

The elements of a communication plan outlined in Figure 6.1 include the following:

- Project purpose/need description—Why is this project needed? What value will it bring to a community? The description of the project purpose and need must be easy to understand and should include clear statements about how it will benefit the community.

- Program/project challenges and opportunities—Begin the outreach plan by analyzing the situation and summarizing potential challenges and opportunities. This allows your team to begin to develop strategies to successfully address challenges and make the most of all opportunities.

- Communication goal and objectives—Why do you want to talk with people in your community or service area about the water reuse project? Objectives describe the results you want to achieve.

- Community research—How will you learn about the interests and opinions of people who live in your community? There are a variety of research techniques to use including one-on-one meetings with community leaders, focus groups, and public opinion research via online or telephone surveys.

- Audiences—Who do you want to talk with? Initially, this list will include broad audience categories such as business groups, environmental organizations, industries, multicultural communities, faith-based groups, the health community, children, students, and so on. A specific list that is reflective of the specific community should be developed as an audience list or database that represents the range of interests in a given community. This mail and e-mail contact list should be updated throughout the life of the project. The communication plan will include sections that identify outreach strategies and activities related to specific audiences (business groups, environmental organizations, academia, health community, etc.). Agency or utility

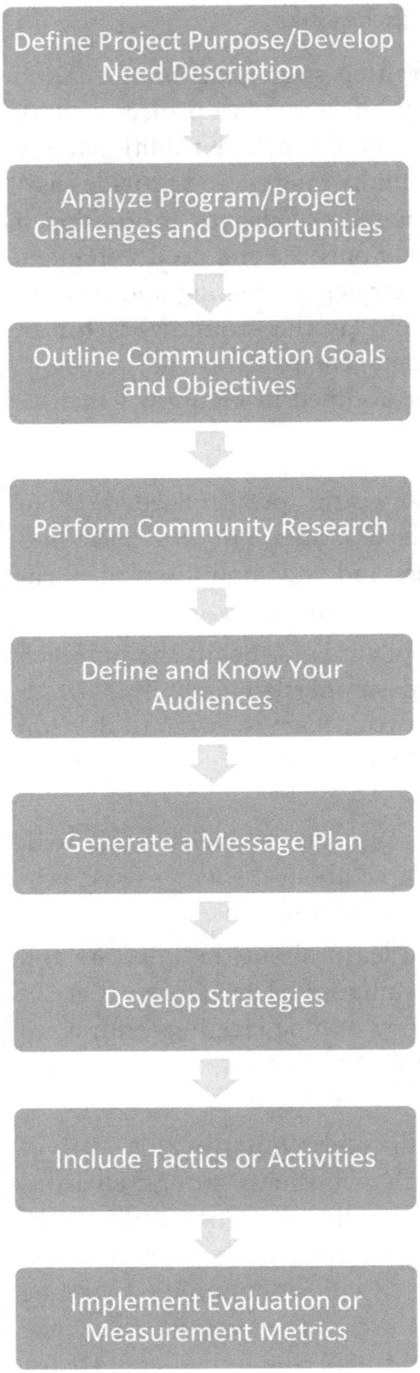

Define Project Purpose/Develop Need Description

Analyze Program/Project Challenges and Opportunities

Outline Communication Goals and Objectives

Perform Community Research

Define and Know Your Audiences

Generate a Message Plan

Develop Strategies

Include Tactics or Activities

Implement Evaluation or Measurement Metrics

FIGURE 6.1 Elements of a communication plan.

staff, decision-makers at all levels, and nonpotable reuse customers are important audiences to include. The media is an audience that will need information about the project and articles in print media or broadcast media stories help the utility inform the broader community about the project so, in that way, the media can also be considered a strategy.

• Messages—What do you want to tell people in the community about the project? These are your messages and they are key to successful communication. Messages should be succinct ideas that are included in all communication, both written and verbal, throughout the life of the program or project. Messages need to be supported by facts. For example, one message for a potable reuse project using full advanced treatment might be, "Potable reuse, or purified water, uses advanced, multi-stage treatment processes to provide a safe, reliable, and sustainable drinking water supply". Try to limit key messages to three, with supporting facts that provide additional information for each message. For example, *Model Communication Plans for Increasing Awareness and Fostering Acceptance of Direct Potable Reuse* (Millan et al., 2015) includes the following three key messages for potable reuse:

  ○ Potable reuse provides a safe, reliable, and sustainable drinking water supply;

  ○ Using advanced purified water is good for the environment; and

  ○ Potable reuse provides a locally controlled, drought-proof water supply.

- Strategies—How will you accomplish your communication goal and objectives? Outreach strategies provide the structure for the activities that will be included to guide implementation of the plan.

- Outreach activities—Outreach activities, typically called *tactics*, comprise a set of communication actions or tools appropriate for each audience that will be used to carry out the identified strategy to meet the goal and objectives. There are many activities that could be included in an outreach plan, but they will likely include the following basics: developing and producing informational materials in a variety of formats including fact sheets, FAQs, infographics, community presentations and information boards; establishing and training staff to participate in a speakers' bureau; participating in community events; providing tours of pilot or demonstration facilities; and more.

- Evaluation or measurement metrics—Keeping track of what you are doing and how members of the community are responding is important as a measure of effectiveness. More importantly, gauging the effectiveness of your outreach program allows you to revise the strategic communication plan to remove activities that do not work and add others that will be more effective or reflect the current phase of the project. For example, as a project moves from design to construction, there may be different concerns or audiences that will require new outreach activities to provide important information and address community concerns.

Once you are ready to actually start the outreach program, adopt the "go to them" vs "come to us" philosophy of community outreach. Community members are busy people, so coming to a meeting in your office to learn about your water reuse project is likely to attract few attendees, if any. Seek opportunities to speak to established community groups that are already on the schedules of your customers; in other words, "go to them". Make a presentation at the chamber of commerce breakfast meeting, attend evening planning group meetings and talk about your project, and participate in community events so you can talk one-on-one with residents. Your audiences are already gathered; they just need you to step up and talk to them.

Agency or utility staff are the best people to speak to community members and the media about the project, but they must be well prepared to do so. Preparation and practice are key to success and should not be optional. Ensure all staff know and can deliver the project key messages comfortably. Workshops will allow staff to practice delivering a standard community presentation and respond to questions that might be asked about the project. In addition, refresher workshops for speakers should be conducted periodically, especially when the project moves into new phases such as design and construction.

Focus initial outreach efforts on community leaders. Just as community members are often too busy to come to a meeting at your office, they may also not want to spend time learning about your project and may prefer to "take the word of" a community leader whose opinion they value. Endorsements from community leaders or on behalf of civic organizations across the broad spectrum of interests in a community will make a difference in the level of support for a project. Leaders who understand the benefits of a local, drought-proof water supply for their community can also be third-party spokespeople about the project to policymakers, media representatives, and their circle of friends and associates. There have been examples of political candidates who have chosen to make potable reuse a campaign issue, so the more factual information community leaders know about a reuse project and its potential benefits for their community, the less damage a negative political campaign may have.

Figure 6.2 illustrates how to work with various audience groups. As discussed above, communication should start with a more targeted and specific audience and become broader to the general public as the outreach plan is implemented.

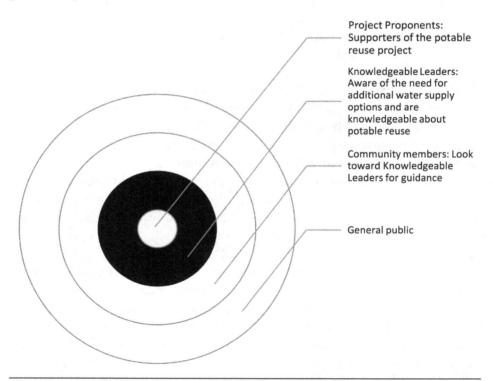

FIGURE 6.2   Target audience (Millan et al., 2015).

An excellent way to expand and extend the reach of your project information and raise awareness more broadly in the community is to provide articles, videos, and graphics to include in the communication channels of community organizations. An article about a local water supply project that appears in the nurses' association newsletter, for example, will garner a vastly different audience than the same article appearing in the local newspaper. That does not mean you ignore the community paper; rather, look for *all* the communication vehicles you can use to reach as many community members as you can.

Many individuals get information about a project through various social media platforms, so having Facebook, Instagram, LinkedIn, and Twitter accounts can allow a utility to reach new audiences. When the Pure Water San Diego Program staff in California purchased Facebook ads, for example, their audience grew and new community members became aware of the project. Social media platforms can also provide a vehicle to receive valuable feedback, publicize facility tours and the availability of speakers for community organizations, or target a specific demographic in the community.

3.0 POTABLE REUSE OUTREACH: TRADITIONAL COMMUNICATION ACTIVITIES AND MUCH MORE

Potable reuse projects must include all the previously mentioned outreach activities and more. Research has shown that there is a high comfort level with recycled water use for irrigation and industrial purposes, but the closer recycled water gets to human contact, the lower the support. *Model Communication Plans for Increasing Awareness and Fostering Acceptance of Direct Potable Reuse* (Millan et al., 2015) includes a report on research conducted in California in 2014 that validates public opinion research by individual utilities that contains this same finding. The most important piece of information your community needs is that your proposed potable reuse project will provide water that is safe to use to augment the community's drinking water supply. The outreach program must be robust, consistent, and sustained to ensure this information is disseminated broadly into the community and reaches the range of interests, ages, genders, and ethnicities to boost awareness and confidence about the project. Although many of the outreach activities outlined in this section for a potable reuse project can also be used for a nonpotable one, water tastings are an example of a very important outreach activity for a potable reuse project that would not be appropriate for a nonpotable one. Also, if the agency or utility has established a good communication structure for its nonpotable reuse projects and public awareness about uses and benefits of reuse is already high, the transition into potable reuse projects will be smoother.

Phase one of the WateReuse Research Foundation's *Model Communication Plans for Increasing Awareness and Fostering Acceptance of Direct Potable Reuse* (Millan et al., 2015) includes a sample community-level outreach plan that can be used for any type of water reuse project, although it is focused on direct potable reuse. The second phase of the project produced sample informational materials that can be easily adapted by any water agency. And while having a plan is a critical first step for any outreach project, the plan must be "worked" day in and day out to break through the information overload (or clutter) we all experience. Consistent and sustained outreach activities allow you to increase community awareness about your specific project.

As such, emphasize the following and refer to Figure 6.3:

• Research—Understanding community opinions is even more important when you talk about potable reuse. Although any community member could have a concern about the safety of potable reuse, whether it involves augmenting a groundwater basin or a surface water reservoir or augmenting the water delivered directly to a drinking water facility or to the water distribution system, concerns will vary among demographic groups. Media coverage about microconstituents, pharmaceuticals, and water quality, in general, can contribute to these concerns. Research methodologies will help to identify the concerns that exist in a specific community so they can be addressed. Begin the research program with one-on-one meetings (sometimes referred to as "in-depth interviews") with community leaders. Ethnic and minority groups that have experienced a variety of factors may be even more skeptical about the safety of potable reuse. These factors include culturally based lack of trust in drinking water safety in a country of origin, belief that members of their community have been treated as "guinea pigs" in the past, experiences of environmental injustice resulting from having their community be the "dumping ground" for unwanted facilities, or feeling they have no voice in decision-making about projects that affect their community. It is, therefore, important to make sure to include leaders in multicultural communities in the one-on-one meeting process. Other ways to continue research into community opinions are to conduct focus groups to test messages, graphic images for informational materials, message carriers, and more. Finally, conducting random public opinion telephone survey research will provide generalizable information about how much members of the community know about potable reuse and the types of information that will provide them with the most confidence that this is a healthy, safe water supply source;

FIGURE 6.3   Potable reuse outreach success factors.

- Internal communication—The agency staff and policymakers are a critically important audience when it comes to potable reuse project implementation. Employees need to learn about the project purpose, treatment process, and water quality just like community members do. Your staff can be your best spokespeople if they are comfortable with potable reuse; after all, they all have family, friends, neighbors, and other acquaintances who will ask them what they think of a project. Provide facility tours and tasting opportunities for employees as soon as possible, before you offer the same opportunities to community members, and make sure all employees and policymakers have the opportunity to ask questions and express their own concerns if they have them. Your staff can also be your worst nightmare if one of them makes a derogatory comment about the project to a community group or says they would be afraid to drink the water;

- Terminology—The water industry has a lot of what most community members would consider jargon. We know what *flocculation* and *tertiary treatment* mean, for example, but most people do not. Use words that community members will understand. Many people can understand that you "clean" the water and that it is "filtered" before use, but they are much less certain about what it means to "treat" the water. Think carefully about the terminology you use to ensure your informational materials, website, presentations and more are going to be understood. There are some good resources to help you do this including the Phase II *Model Communication Plans for Increasing Awareness and Fostering Acceptance of Direct Potable Reuse* (Millan et al., 2015) and *Talking About Water: Vocabulary and Images that Support Informed Decisions about Water Recycling and Desalination* (Macpherson, 2011);

- Technology descriptions—When it comes to potable reuse, it is okay to "talk tech" as long as descriptions are easy to understand and use layperson terms as much as possible. When potable reuse includes advanced technological processes such as microfiltration, reverse osmosis, and UV light disinfection, relating other uses for these processes can build confidence. For example, when describing reverse osmosis you might add the statement, "This is the same process used by some bottled water companies and baby food manufacturers". Likewise, relating the size of the pores in microfiltration fibers as being "3,000 times smaller than the width of a human hair" helps people understand that not many particles will pass through the system;

- Graphics and videos—Graphics and videos that show potable reuse processes reinforce not only the meticulously planned nature of the project, but help make the words used in informational materials come alive to people. Very brief videos can underscore what is happening to make the water produced in the potable reuse project safe to drink, and animations can also show what happens inside the complex advanced technical processes like UV light disinfection. The WateReuse Association website, watereuse.org, is a good source of graphics and videos, including video interviews with experts in the reuse and water quality fields;

- Alliances and partnerships—Focusing outreach efforts on community leaders and the organizations they are part of is an extremely valuable outreach strategy for potable reuse projects. Support from partners and third-party allies allows for increased dissemination of project information through the communication channels of these external groups and raises awareness about the process and a specific project. Forming partnerships with strong community groups, civic organizations such as Rotary or Optimists, universities, and interest groups ranging from environmental organizations to the health community to faith-based groups can pay big dividends in terms of showcasing different aspects and benefits of potable reuse at forums and other gatherings;

- Youth outreach—Adding potable reuse content to existing school education programs will mean that youth activities pages and very basic technological process descriptions must be developed. This helps you talk tech in understandable terms and allows you to lay groundwork with your future customers, not to mention their parents and teachers. El Paso Water's TecH2O, a water resource learning center, and the education center at the LOTT Clean Water Alliance (of Texas and Olympia, Washington, respectively) incorporate displays and other learning tools that can enhance the educational experience of local schools. Pure Water San Diego staff has also developed activity pages for school groups, scouting programs, and other young children that tour the Pure Water demonstration facility. These programs and many other utilities have robust school education programs that they are willing to share with others;

- Facility tours/water tastings—The gold standard for a potable reuse project is to invite community members to tour the facility to see for themselves the quality of the finished water; furthermore, being able to taste the product converts most people to a high level of comfort with potable reuse. Orange County Water District's Groundwater Replenishment System staff conduct a pre- and post-tour survey and have found that 97% of visitors are comfortable with having "advanced purified/recycled water" as part of their drinking water, compared with a much lower percentage who either did not know if it was safe or actually thought it was not safe to drink before they toured; and

- Rapid response plan—With the advent of social media and the associated extremely fast sharing of information today, it is important to correct misinformation rapidly to minimize damage to the project and agency reputation. Planning how to respond to misinformation or any other type of crisis long before it happens provides for an orderly approach to relaying accurate information to the community. Monitoring blogs and websites where comments and critiques could be posted will also allow for prompt responses and the ability to address errors rapidly.

4.0 REFERENCES

Millan, M.; Tennyson, P.; Snyder, S. (2015) *Model Communication Plans for Increasing Awareness and Fostering Acceptance of Direct Potable Reuse*; WRRF 13-02; WateReuse Research Foundation and Metropolitan Water District of Southern California: Alexandria, Virginia.

Macpherson, L. (2011) *Talking About Water: Vocabulary and Images that Support Informed Decisions about Water Recycling and Desalination*; WRRF 7-13; WateReuse Research Foundation: Alexandria, Virginia.

National Academy of Sciences (2012) *Water Reuse: Potential for Expanding the Nation's Water Supply through Reuse of Municipal Wastewater*; The National Academies Press: Washington, D.C.

5.0 SUGGESTED READING

WateReuse Foundation (2006) *Marketing Nonpotable Recycled Water: A Guidebook for Successful Public Outreach & Customer Marketing*; WateReuse Foundation: Alexandria, Virginia.

7

Implementation: Treatment Technologies and Other Project Elements

Larry Schimmoller, P.E.; Jason Assouline, P.E.; and Tyler Nading, P.E.

The purpose of this chapter is to review the implementation of water reuse projects, specifically focusing on treatment technologies that can reliably meet water quality criteria requirements and achieve the identified treatment goals. Additional topics relevant for implementation of water reuse projects are discussed in this chapter, including treatment process sizing, operability, staffing, blending, and source control, among others.

Treatment process selection for water reuse should be driven by the quality of the source water, the end use of the water, and the regulatory requirements. As discussed in Chapter 1, *water reuse* can refer to a wide range of treated water quality. When discussing water reuse, it is important to acknowledge and distinguish between nonpotable and potable reuse and the specific considerations necessary when implementing either form of water reuse. This chapter will discuss the treatment drivers and approaches for each of these different types of water reuse.

1.0   NONPOTABLE REUSE

Nonpotable reuse is the most common and widely accepted type of water reuse because it has been implemented in a variety of applications worldwide as a means of conserving potable water resources by providing water for irrigation, industrial and commercial applications, and other uses (e.g., recreational enhancements). Because of the diversity in end uses for nonpotable reuse, the treatment goals and, subsequently, the treatment processes implemented to achieve the desired water quality are based on the specific nonpotable application requirements in addition to the requirements established by reuse regulations to ensure protection of public health.

The treatment requirements for industrial reuse are typically much different than irrigation application as the water quality goals are focused on the needs of the end user and the industrial process where the water is to be used. Nutrient (e.g., nitrogen, phosphorus) and ion removal (e.g., chloride, hardness) can be required to meet industrial water quality requirements for cooling and boiler water applications, although these treatment requirements are very site-specific. The economic considerations for implementing reuse for an industrial application can also be fairly complicated depending on each specific situation. Considerations include cost savings from not having to develop a new potable water supply, cost comparison of potable and nonpotable water, and evaluating if more nonpotable water is required than potable water because of potentially lower water quality and reduced cycles of concentration.

1.1   Agricultural and Irrigation Reuse

Utilities consider nonpotable reuse as a means to offset the use of potable water and freshwater resources for large and often seasonal demands that do not require potable quality water (i.e., parks and golf course irrigation, agricultural irrigation). Although most agricultural irrigation is typically supplied by untreated water (surface or groundwater), nonpotable reuse water may also be used for irrigation of either edible or nonedible crops. The approved uses for nonpotable water are state-specific and can vary significantly between states.

Because water resource recovery facilities (WRRFs) are focused on providing treatment to meet surface discharge regulations, the nonpotable reuse supply requires additional treatment to address the increased potential for human contact in water reuse applications. Refer to Chapter 4, "Regulations and Risk Assessment", for additional discussion on the regulatory drivers for nonpotable reuse treatment.

1.1.1   Water Quality Requirements and Treatment Goals

Nonpotable reuse treatment requirements are typically more stringent than the existing WRRF discharge resulting in the use of tertiary treatment process(es) after secondary treatment. Regulations for nonpotable reuse typically target a reduction in total suspended solids (TSS) and pathogens to ensure safety given the potential for human contact with the reuse water; this often results in the need for tertiary filtration and high-level disinfection of secondary effluent. Although treatment and regulations are not typically focused on inorganic constituents in the reuse water, special considerations for the ionic makeup of the water, including parameters such as boron, chloride, total dissolved solids (TDS), and the sodium adsorption ratio, can influence the viability of utilizing nonpotable reuse as an irrigation source.

1.1.2   Treatment Approaches

Typical treatment for nonpotable reuse includes tertiary filtration and disinfection to meet the requirements for the specific reuse application. Filtration to remove suspended solids and pathogens may be achieved with a number of technologies, including granular media filters, moving bed sand filters, cloth filters, and membrane filters. The state of California maintains a list of approved filtration technologies for use in nonpotable reuse applications that is useful for those considering tertiary filtration (California Title 22). Primary disinfection is typically achieved with chlorine contact or UV radiation, and some nonpotable reuse systems maintain a chlorine residual in the distribution system for secondary disinfection and biogrowth control.

1.2   Industrial Reuse

Industrial reuse for nonpotable application is beneficial in applications where large quantities of water may be used and potable water quality is not necessary for the process. The project drivers for industrial uses are typically the economic considerations associated with growing water scarcity. Common industrial applications for industrial nonpotable reuse include boiler feed and cooling towers. Source water for industrial reuse can either be from a municipal secondary effluent source or from the discharge from the industrial user itself.

Treatment requirements vary based on the specific intended use; however, considerations for scaling and corrosion in boiler feed and cooling towers typically include water quality limits for chloride, TDS, ammonia, TSS, and bacteria. Treatment approaches also vary based on the water quality requirements specified by the industrial end user. Based on the requirements of the system, no additional treatment beyond the tertiary nonpotable treatment system may be necessary or a stand-alone advanced system may be necessary to create high-purity water from the nonpotable reuse supply. If an advanced treatment system is needed, this is typically added at the point of use by the industrial user.

1.3   Implementation of Nonpotable Reuse

Nonpotable reuse is relatively widespread and common in states that have limited water supplies. Significant efforts have been made to educate the public about water usage in these areas. Refer to Chapter 6 for additional discussion on communication and outreach for nonpotable reuse systems.

Nonpotable reuse systems are typically implemented as tertiary treatment infrastructure added at the WRRF because co-located treatment infrastructure typically presents the lowest operational cost. However, nonpotable treatment systems may also be completely stand-alone WRRFs. In addition to the treatment infrastructure, nonpotable reuse systems require a stand-alone distribution system (also known as a "purple pipe system"), which directly supplies the water to the various end users. The dual distribution system necessary for nonpotable reuse can be a barrier to implementation because the additional infrastructure required to supply the customers (and subsequently maintain and add to the system for new users) may be cost-prohibitive (NRC, 2011). Satellite facilities or treatment located at the end user's facility can avoid the need for a dual distribution system in certain situations.

1.3.1   Sizing and Operability

Sizing treatment and storage facilities for irrigation or agricultural reuse can be difficult and/or costly given the seasonal and/or diurnal demands for the reuse water. These facilities can be costly, and can cause operational difficulties in starting and stopping the reuse treatment processes. Some utilities have identified alternative end uses of the reuse water such as equalization tanks, reservoirs, or aquifers, so that the treatment process can maintain a consistent flowrate. Others elect to base-load the demand with reuse water and use alternate sources to account for demand variations.

Similar to conventional WRRFs, the nonpotable reuse system is also often sized to be able to add future customers. This can result in a system that operates at less than full capacity in the short-term and has negative consequences such as water age, loss of residual, and biological regrowth. Conversely, industrial reuse is typically baseloaded with minimal variation in the reuse water demand.

The operation of facilities for irrigation and agricultural use, typically only filtration and disinfection, does not add any specific challenges because these treatment methods are well understood. Operation of facilities for industrial users varies in complexity and requires training specific to the selected processes.

1.3.2 Staffing

Although there is currently not a specific operator license designation for operating a water reuse facility (either nonpotable or potable), studies focusing on special licensure, training, or certifications are necessary to address the additional operational considerations associated with successfully implementing advanced reuse treatment (particularly for potable reuse). A facility may be required to provide additional or specific staffing based on regulatory or service-level requirements. Nonpotable reuse typically consists of tertiary treatment at the WRRF and, therefore, any additional staffing falls under the purview of the WRRF operations staff. Stand-alone advanced treatment for industrial users is typically staffed by the end user or the utility providing the treatment.

2.0 POTABLE REUSE

Potable reuse is the practice of using treated wastewater as a drinking water source and typically incorporates advanced treatment processes to remove organics, pathogens, and other contaminants to meet potable water standards. Similar to the distinction between nonpotable and potable reuse, it is important to distinguish between different types of potable reuse. As discussed in Chapter 1, indirect potable reuse (IPR) refers to a system in which WRRF effluent or advanced treated effluent is introduced to an environmental buffer before being withdrawn for potable purposes (WateReuse Foundation, 2015). Direct potable reuse (DPR) refers to a system that has no environmental buffer between WRRF effluent and potable water, although blending strategies can be implemented and still considered to be DPR. Figure 7.1 provides common approaches to potable reuse.

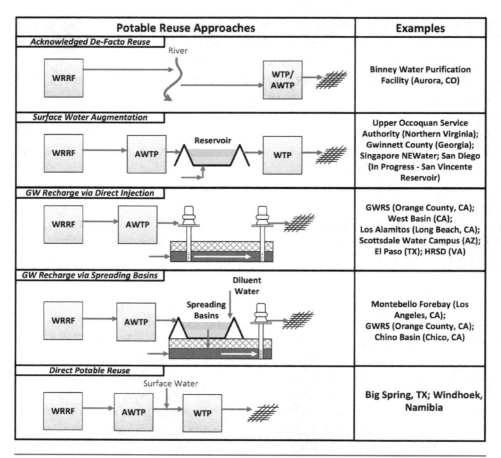

FIGURE 7.1 Common approaches to potable reuse.

A central theme to implementation of potable reuse projects, both indirect and direct, is the development of multiple barriers of treatment against organics and pathogens. *Multiple barriers of treatment* refer to separate treatment processes in series targeting specific contaminants or classes of compounds so that treatment (removal, destruction, transformation, or inactivation) can be achieved at multiple locations and that failure of a single barrier does not result in a failure of the entire treatment system (WateReuse Foundation et al., 2015). Treatment processes can be redundant in the removal of some compounds (microfiltration and reverse osmosis both provide a pathogen barrier) and complementary in the removal of others (reverse osmosis is an organics barrier, but not microfiltration) (Schimmoller and Kealy, 2014). The use of multiple barriers achieves a higher level of system reliability and resilience. Different types of treatment barriers are discussed in the following section.

2.1   Water Quality Requirements and Treatment Goals

As discussed in Chapter 4, there are currently no federal regulations for potable reuse projects. Operational potable reuse facilities have been subject to achieving different standards of treatment as required by state, local, or project-specific requirements. However, treatment requirements across potable reuse projects commonly include similar parameters and goals. Table 7.1 provides treatment requirements for representative projects.

Potable reuse projects also often have separate, nonregulatory treatment goals or performance indicators. Treatment goals are implemented to achieve superior treatment relative to the regulatory requirements and are supported by operational protocols often referred to as *critical control points* (CCPs). Performance indicators refer to nonregulatory parameters such as emerging contaminants that are not currently regulated, but that are often detectable in WRRF effluent and can be used to indicate the treatment performance of the water reuse facility (i.e., breakthrough of a specific compound could indicate the need for replacement or regeneration of granulated activated carbon [GAC]).

In general, the treatment goals between IPR and DPR are very similar and focus on multiple treatment barriers for pathogens and organics. However, because DPR lacks an environmental barrier that provides dilution, natural treatment, and response time, DPR can require more safeguards than IPR. This can include implementation of CCPs, additional treatment barriers, and/or automatic diversions when the finished water does not meet treatment requirements. Refer to Chapter 8 of this publication, *Critical Control Point Assessment to Quantify Robustness and Reliability of Multiple Treatment Barriers of a DPR Scheme* (WE&RF, 2016a), and *Guidelines for Engineered Storage Systems* (WE&RF, 2016b).

2.2   Treatment Approaches

A variety of treatment trains have proven successful in potable reuse projects using an array of different unit treatment processes. Advanced potable reuse treatment trains may be membrane-based (i.e., microfiltration/ultrafiltration [MF/UF], reverse osmosis, UV advanced oxidation process [UVAOP], also known as *full advanced treatment* [FAT] in California) or non-membrane-based (i.e., biologically active carbon [BAC] filtration and/or GAC combined with other treatment barriers). Soil aquifer treatment, which can provide an excellent barrier to pathogens and organics, may be coupled with membrane or non-membrane-based treatment as part of the treatment process. All treatment approaches should include a multiple barrier approach for pathogens, bulk organics, and nonregulated trace organics (including pharmaceuticals and personal care products).

TABLE 7.1   Water quality requirements for representative potable reuse projects.

Project/ location	UOSA (Virginia)	OCWD groundwater replenishment system (California)	Montebello Forebay (California)	Big Spring (Texas)
Type	Surface water augmentation	Groundwater injection	Groundwater recharge via SAT	Direct potable reuse
Maximum contaminant levels (MCLs)	—	Comply with all primary drinking water MCLs		
Nitrogen	Total Kjeldahl nitrogen < 1 mg/L	Total nitrogen < 10 mg/L	Total nitrogen < 10 mg/L[a]	NO_3-N < 10 mg/L; NO_2-N < 1 mg/L
Solids	TSS < 1 mg/L; turbidity < 0.5 NTU	Turbidity < 2 NTU	Turbidity < 2 NTU	—
Organics (TOC/ carbonaceous oxygen demand [COD])	COD = 10 mg/L (~ 3.8 mg/L TOC)	TOC < 0.5 mg/L	TOC < 0.5 mg/L/RWC[b]	—
Enteric viruses[d]	Multiple barriers required (total coliform < 2/100 mL)	12-log LRV	12-log LRV[c]	8-log LRV
Cryptosporidium[d]		10-log LRV	10-log LRV[c]	5.5-log LRV
Giardia[d]		10-log LRV	10-log LRV[c]	6-log LRV
Minimum treatment requirements	UOSA treatment process	Oxidation, filtration, disinfection, reverse osmosis, AOP	Oxidation, filtration, disinfection, SAT	Microfiltration, reverse osmosis, UVAOP

[a]Sample may be taken before or after surface application.

[b]See California regulations for surface spreading to calculate TOC requirement given the recycled water contribution (RWC).

[c]1-log virus reduction is credited in California for each month the recharge water is retained underground; 10-log Cryptosporidium and 10-log Giardia reduction is credited in California if the recharge water is retained at least 6 months underground.

[d]Pathogen removal credits are measured from raw wastewater through advanced treatment for California projects; Big Spring pathogen removal credits are measured from secondary effluent through advanced treatment.

Understanding the quality of the potable reuse source water is important to developing the treatment approach. If the source water, like most potable reuse projects, is WRRF secondary effluent (final effluent is typically not used because of disinfection byproducts formed if the WRRF uses chlorine for its disinfection process), the role of the WRRF needs to be understood. Water resource recovery facilities can provide an excellent barrier of treatment for pathogens, organics, and trace contaminants. Water resource recovery facilities provide good removal of pathogens, and WRRFs with longer solids retention times, typically those that use biological nutrient removal, provide significantly better removal of organics and are preferred in potable reuse applications. This is particularly important for non-membrane-based potable reuse approaches that have limited nitrogen removal. As mentioned in endnote "d" of Table 7.1, the treatment performance of the WRRF can provide credit toward the treatment requirements of the potable reuse system. As potable reuse becomes more prevalent, many utilities are considering treatment technologies such as a membrane bioreactor (MBR), which bridges the gap between a WRRF and reuse and provides benefits for both systems. It is also critical to understand and account for WRRF water quality process excursions and diurnal fluctuations (i.e., variable nitrogen load) at the WRRF because these can provide difficulty for the advanced treatment processes.

2.2.1 Membrane-Based Treatment

Membrane-based treatment trains for potable reuse typically include microfiltration or ultrafiltration for removal of particles, nanofiltration or reverse osmosis for removal of dissolved salts, and UV) disinfection or UVAOP. This is commonly referred to as *FAT* because this combination of treatment processes has been most commonly implemented in potable reuse projects in California. California's regulatory requirements require FAT for potable reuse projects that practice direct injection to a potable aquifer and are considering its use in DPR projects. An example process flow diagram (PFD) of this treatment train is shown in Figure 7.2.

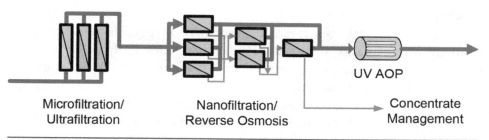

FIGURE 7.2 Simplified PFD of membrane-based potable reuse treatment train.

A key advantage of membrane-based treatment trains is in the reduction in TDS and total organic carbon (TOC) provided by nanofiltration/reverse osmosis treatment. These processes also provide an excellent pathogen barrier, although log-removal credits are currently limited because of the ability of online analyzers to reliably measure removal. Disadvantages of the membrane-based treatment trains include relatively high capital and operating costs of nanofiltration/reverse osmosis and the management of membrane reject or concentrate, especially for inland locations where ocean disposal is not an option (Schimmoller and Kealy, 2014). Figure 7.2 provides a common example of a membrane-based treatment train, but other trains are also viable, including MBR-reverse osmosis, which is discussed in Section 2.5.4.

2.2.2 Non-Membrane-Based Treatment

Potable reuse treatment trains that do not include nanofiltration/reverse osmosis may use a combination of different treatment processes to achieve multiple barriers of treatment. There is a wide array of operational potable reuse treatment processes, and although it can be considered that no two treatment approaches are exactly the same, there are common threads between non-membrane-based schemes, specifically including a backbone organics removal treatment process consisting of either GAC, soil aquifer treatment (SAT), or a combination of both. It should be noted that non-membrane-based trains do not remove dissolved salts from the source water, so it is only applicable if the dissolved salt concentration meets finished water quality requirements or other water is available for blending to meet TDS goals. Additionally, the potential for salt accumulation within the reuse system should be evaluated to ensure that there is enough alternative source water to negate the effect of salt buildup.

Pretreatment of WRRF secondary effluent is typically included in a non-membrane potable reuse process. This can include lime clarification or metal salt coagulation with conventional sedimentation or high-rate clarification (i.e., sand-ballasted clarification, solids contact clarification). The goal of pretreatment is to reduce the solids load, remove organics, and provide security against WRRF excursions.

Particle and pathogen removal is required in potable reuse to meet drinking water quality standards. This is most commonly achieved with granular media or membrane filtration operated to meet drinking water turbidity standards. Granular media filtration can provide enhanced organics removal when operated in a biological filtration mode, provided upstream oxidation is practiced and proper media selection is made to support biological growth. Ion exchange, although not considered a conventional particle removal process, can be used for the targeted removal of specific compounds (i.e., ions or organics).

Adsorption of organics is commonly used in non-membrane-based approaches. This is typically achieved using GAC reactors in which the carbon is replaced with regular frequency to maintain the adsorption capacity. Frequency of carbon regeneration dictates the viability of using GAC. Operation of GAC reactors in a biological mode can enhance organics removal and reduce GAC replacement requirements. Powdered activated carbon can also be used to achieve similar organics adsorption.

Lastly, an advanced oxidation process that generates the hydroxyl radical is typical for potable reuse to provide destruction of recalcitrant and trace organics. Advanced oxidation is typically provided by ozone and hydrogen peroxide or as part of UVAOP. The UVAOP process combines advanced oxidation and photolysis, but smaller doses of UV radiation can also be used for photolysis only, which can provide an excellent pathogen barrier but provides limited oxidation of organics. This is referred to as *UV disinfection*.

Although it is acknowledged that there is no "typical" non-membraned potable reuse treatment train, Figure 7.3 shows an example of how the unit processes previously described can be assembled into a treatment process that provides multiple barriers of treatment.

2.2.3 Soil Aquifer Treatment

Soil aquifer treatment refers to natural treatment that occurs as water travels through soil or an aquifer and is a component of an IPR treatment scheme. The feed water can be distributed by surface spreading and allowing the water to percolate through the vadose zone to the aquifer. Direct injection to the aquifer can also achieve treatment as the water travels through the aquifer system. Soil aquifer treatment has proven to be an excellent barrier for nutrients, pathogens, and organics and, depending on the system, only minimal additional treatment may be required. The use of SAT is highly dependent on local geologic conditions.

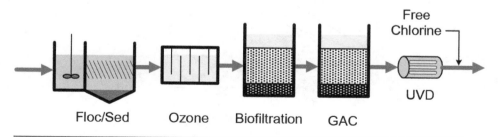

FIGURE 7.3   Simplified PFD of non-membrane-based potable reuse treatment train.

2.2.4 Relative Cost of Potable Reuse

If both membrane and non-membrane treatment trains are able to achieve the treatment requirements, cost considerations often become the driver for the selection of the treatment train. Figure 7.4 shows relative costs (in 2014 dollars) for different membrane and non-membrane potable reuse schemes. Costs are high-level, but show the magnitude of difference between different processes and highlight the importance of concentrate management in the membrane-based scheme.

2.3   Implementation

An important aspect in the implementation of a potable reuse project is addressing the public perception of the project, which is discussed in Chapter 6. This section will focus on other key implementation factors that can help projects be successful.

2.3.1 Water Resource Recovery Facility Pretreatment Control Program

Water resource recovery facilities are required to enforce a pretreatment control program by U.S. Environmental Protection Agency regulation 40 CFR 403, General Pretreatment Regulations for Existing and New Sources of Pollution. These regulations are meant to protect the WRRF influent quality

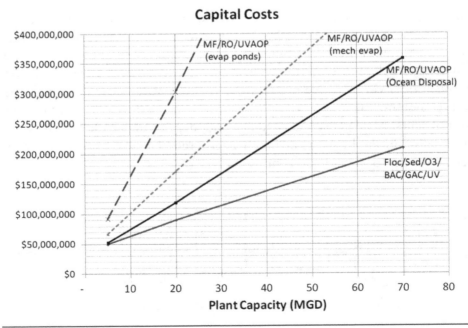

FIGURE 7.4   Costs of implementing potable reuse treatment schemes (mgd ÷ 0.2642 = ML/d) (adapted from Schimmoller and Kealy, 2014).

from heavy commercial and industrial discharges. When considering implementation of a potable reuse system, the pretreatment control program should be re-evaluated to include the following:

- Sampling of heavy industrial users to determine if local limits should be made stricter or if the discharge should be diverted away from the potable reuse system and
- Sampling at the WRRF influent and secondary effluent to determine if there are recalcitrant contaminants that may not be removed through the conventional or potable reuse processes and subsequent removal of the source of the contaminant.

2.3.2 Sizing and Operability

Water resource recovery facilities are sized to treat both the diurnal dry weather flows and the peak wet weather flows. Although some potable reuse facilities are designed to treat diurnal flows, they are ideally designed and operated at a more consistent flowrate to provide better treatment and excellent finished water quality. A potable reuse facility may be located at the WRRF or elsewhere, but should be sized based on the potable water demands. When coupled together, the difference in flow profiles can provide operational difficulties.

Depending on the capacity of the potable reuse facility, either a surplus or a deficit of secondary effluent may be available, resulting in either an alternate discharge from the WRRF or the need to augment the overall potable water source to meet the ultimate potable water system demands. An evaluation at the WRRF is required to ensure that the final disinfection and outfall are sized for the range of new design flows.

Equalization within the WRRF process (typically either after primary or secondary treatment) is a strategy that provides attenuation of the diurnal flow fluctuations, can improve treatment at the WRRF, and provides a more consistent feed supply to the potable reuse treatment facility, resulting in better operation and higher finished water quality.

Regardless of the approach to sizing the potable reuse facility, consideration should be provided to the handling of WRRF water quality excursions. If the WRRF is producing secondary effluent that poses problems to the potable reuse facility, the flow should be diverted to the WRRF final disinfection and the outfall.

2.3.3 Blending

Blending is a common element in any potable reuse scheme. Blending with other waters can occur in the ground, river, or reservoir (lake), or, in the case of DPR, with raw water entering a water treatment facility or finished water in the distribution system (see Figure 7.1). Blending can offer benefits to finished water quality (e.g., lower TDS) as well as contribute to the public acceptance of potable reuse water. When embarking on a potable reuse project, the water quality from each source must be carefully analyzed to understand how the various sources contribute to the blended water quality.

2.3.4 Small-Scale Testing

Understanding the treatability of the WRRF effluent at the potable reuse facility is a critical factor in mitigating project risk by evaluating the efficacy of and fine-tuning each treatment process. Bench-scale treatability testing and pilot testing can provide confidence that the selected treatment process will achieve the treatment goals without creating unintended byproducts or treatment consequences. The bench or pilot testing can provide important design criteria information for the sizing of the full-scale system and it can provide additional water quality information that will support the approach to blending and finished water quality. If possible, pilot testing that is operational through multiple seasons provides insight to how the treatment process can sustainably meet the treatment goals throughout the year and significant data on designing and operating the full-scale facility. In the absence of bench or pilot testing, modeling can be performed to predict treatment performance, although it will not provide specific information on treatability that must be achieved through testing.

High-profile potable reuse facilities that may be especially sensitive to the public or that are navigating new regulatory requirements may consider the value of constructing a demonstration facility that shows, at a meaningful scale, that the treatment approach will be successful and allows for better public understanding through educational tours. The size of the facility is relative to the intended full-scale system, but a 2- to 4-MLD (0.5- to 1.0-mgd) facility would be reasonable for a large potable reuse program. Operational data from the demonstration facility can be used to develop the regulatory permitting required for the full-scale system.

2.3.5 Staffing

Development of Operation and Maintenance Plan and Training and Certification Framework for Direct Potable Reuse Systems (Walker, 2017) is a study that focuses on the considerations associated with operations and maintenance planning and training necessary for DPR systems. Although specific regulations are currently not in place, operation of a potable reuse facility does require a higher level of operator and maintenance staff skills based on the advanced treatment commonly implemented at these facilities.

That said, it is important to note that all operations and maintenance staff at any facility (WRRF or water treatment facility) are an essential part of meeting treatment goals and protecting the human health and the environment; specific intricacies of any treatment facility should be managed by qualified staff that have developed site-specific standard and emergency operating procedures as well as routine staff training. It is challenging to delineate potable reuse qualifications or operator standards because of the variability in where the advanced treatment is implemented, what the treatment entails, and how the various sources are blended in the distribution system.

2.4   Case Studies

There are some potable reuse facilities in operation that demonstrate the variability in system components and processes based on the site-specific needs to accomplish the goal of providing a potable water source.

2.4.1   Acknowledged De Facto Reuse

The concept of de facto IPR is common because of the fact that many surface water treatment facilities are downstream of a WRRF outfall (Rice et al., 2013; Rice and Westerhoff, 2014). The Binney Water Purification Facility (BWPF) is a unique potable reuse treatment facility that treats two distinct source waters through separate treatment trains that are subsequently blended at a minimum blend ratio of 2:1 (conventional to advanced) before disinfection. The potable reuse train is supplied by an effluent-dominated river, followed by riverbank filtration and soil aquifer treatment before advanced treatment at the treatment facility. The advanced treatment includes chemical softening, UVAOP, biological filtration, GAC adsorption, and chlorine disinfection; a simplified PFD is shown in Figure 7.5. The conventional train treats a surface water supply through flocculation/sedimentation and biological filtration. This facility has been in operation since 2010.

2.4.2   Surface Water Augmentation

Surface water augmentation projects include advanced treatment at the WRRF, which is subsequently discharged to a water storage reservoir that

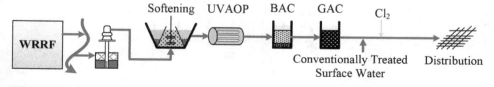

FIGURE 7.5   Simplified PFD of Aurora, Colorado, BWPF advanced treatment train.

receives contributions from other natural water influent sources. Specific utilities that are practicing this potable reuse approach include the Upper Occoquan Service Authority (UOSA, in Virginia), Singapore Public Utilities Board (Singapore PUB), Gwinnett County (in Georgia), and the Western Corridor Recycled Water Project (Queensland, Australia). The advanced treatment at these utilities varies between membrane-based treatment (Singapore PUB and Western Corridor) and non-membrane-based treatment trains (UOSA and Gwinnett County). The UOSA has been in operation since 1978 and the Singapore PUB NeWater facility has been in operation since 2000. The Western Corridor Recycled Water Project, which consists of three potable reuse facilities, was operational for several years around 2010, but has been temporarily shut down because of significant rainfall in the area. A simplified PFD of the UOSA treatment process is shown in Figure 7.6.

2.4.3 Groundwater Injection

The Orange County Water District (OCWD) and the Hampton Roads Sanitation District (HRSD) are based on an IPR process of advanced treatment at the WRRF followed by groundwater injection. The direct injection of the advanced treated water is driven by the site-specific hydrogeology as well as the objective of the groundwater storage and recovery component of the process. The aquifer serves an additional treatment barrier for pathogens and organics and, in the case of HRSD, also provides additional environmental benefits (nutrient reductions for an ecologically sensitive area, seawater intrusion barrier, and land subsidence mitigation). The HRSD has completed piloting and is currently (as of 2017) in the process of constructing a 3.8-MLD (1-mgd) demonstration facility. The OCWD Groundwater Replenishment System (GWRS) has been in operation since 2008, and a simplified PFD is shown in Figure 7.7.

2.4.4 Groundwater Recharge via Spreading Basins

The Montebello Forebay Groundwater Recharge Project (MFGRP) is located in Los Angeles County, California, and has been in operation since

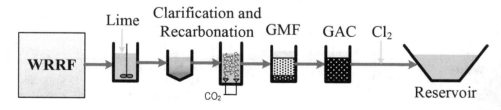

FIGURE 7.6 Simplified PFD of UOSA regional water reclamation facility.

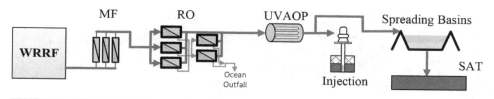

FIGURE 7.7   Simplified PFD of OCWD GWRS.

1962. The MFGRP consists of tertiary treatment (filtration and disinfection) at the WRRF and uses spreading basins and SAT to meet the treatment requirements. Significant treatment of nutrients, pathogens, and organics has been demonstrated via SAT, which allows the MFGRP to operate without advanced treatment facilities. A simplified PFD of the treatment process is shown in Figure 7.8.

2.4.5 Direct Potable Reuse (Big Spring, Windhoek)

The two most established DPR projects in the world are the New Goreangab Water Reclamation Plant (in Windhoek, Namibia) and Big Spring (Texas). Windhoek has been practicing DPR since 1968 with the Old Goreangab Water Reclamation Plant, which was decommissioned in the 1990s. The new facility was put into operation in mid-2002 (Lahnsteiner and Lempert, 2007). The treatment mostly uses a series of non-membrane-based treatment technologies (ozone/BAC/GAC); however, ultrafiltration membranes are also included as an additional barrier at the end of the treatment process.

Big Spring was designed and commissioned in response to an extreme drought in central Texas and the treatment approach includes the microfiltration/reverse osmosis/UVAOP treatment scheme. The treated water from the advanced potable reuse treatment facility is currently blended with a surface water and re-treated through a surface water treatment facility before entering the potable distribution system (the maximum permitted potable reuse contribution is 20%). Note that there is no environmental buffer in this system. A simplified PFD is shown in Figure 7.9.

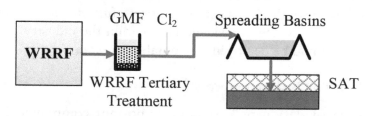

FIGURE 7.8   Simplified PFD of Montebello Forebay Project.

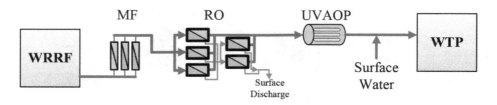

FIGURE 7.9   Simplified PFD of Big Spring regional water reclamation project.

2.5   Innovation and Research and Development

Water reuse, particularly potable reuse, is an emerging segment in the water industry that is garnering a renewed focus as available water resources become less available. Fundamental in the equation of whether potable reuse can safely and cost-effectively provide sustainable potable water supplies for communities is the ability of the water industry to leverage innovation and new technologies to efficiently meet water quality treatment goals. This section briefly describes recent innovations and ongoing research and development efforts applicable to potable reuse.

2.5.1   Ozone, Biologically Active Carbon, and Granulated Activated Carbon

Coupling ozone, biofiltration, and GAC adsorption is a treatment scheme (analogous to microfiltration/reverse osmosis/UVAOP) that has shown to be an effective multibarrier treatment that addresses many of the treatment challenges associated with potable reuse. Generally, preceded by a pretreatment step (i.e., flocculation/sedimentation) and followed by disinfection, this core treatment combination has proven at pilot and full scale to be effective at removing bulk organics, trace organics, and some pathogens. The primary benefit of this treatment scheme is that, for inland utilities, there is no brine waste stream to dispose of so the capital and life cycle cost associated with implementing this treatment is much less than the energy-intensive FAT. Moreover, because these processes are well established (on an individual basis) for drinking water treatment, the equipment is readily available and is relatively easy to operate and maintain. There are numerous projects that are in various stages of study, piloting, demonstration, or full-scale implementation that are gathering a multitude of useful information and data that will contribute to the knowledge base within the industry as to how this approach should be deployed in a potable reuse application.

2.5.2   Advanced Oxidation Process

The advanced oxidation process is an important component in many advanced treatment and potable reuse projects because it serves as a broad barrier for many pathogens, trace organics, pharmaceuticals, and personal care products. Creation of the hydroxyl radical as an oxidant distinguishes the advanced oxidation process (AOP), which is typically achieved using either hydrogen peroxide and ozone or by using UV reactors for UVAOP. The UVAOP is most commonly achieved with the addition of hydrogen peroxide to form hydroxyl radicals upstream of a high-dose UV system. Ultraviolet/chlorine (UV/HOCl or UV/ClO$_2$) AOP treatment is a relatively newer approach in AOP technologies that offers the potential to provide an economical and effective treatment barrier. Low pH ($\sim$6.0) is critical for effective operation of the UV/HOCl approach.

Hydrogen peroxide is an excellent source for generating hydroxyl radicals; however, it is an expensive and potentially hazardous chemical. Advantages that have been demonstrated for UV/chlorine AOP include the following: effective trace organic oxidation, lower energy consumption required for the UV system, lower chemical costs, and low disinfection byproduct (halogenated organic and N-Nitrosodimethylamine [NDMA]) formation (Rosenfeldt et al., 2013; Sichel et al., 2011). Using hydrogen peroxide and ozone for AOP has proven successful in some applications, but requires consideration for bromate and NDMA formation and does not achieve the same pathogen reduction as UVAOP.

2.5.3 Concentrate Management

Concentrate (i.e., brine) management is a critical component in evaluating whether high-pressure membranes are a viable treatment technology to be used in a potable reuse process. At coastal facilities that have the ability to dispose of this waste stream into a highly saline environment (i.e., ocean or sea), the life cycle cost compared to the treatment efficacy often proves to be economically viable. At inland utilities where deep well injection is not an option, it is often cost-prohibitive (if not technically infeasible) to dispose of the concentrate (Schimmoller and Kealy, 2014) via evaporation ponds, brine concentration, or mechanical evaporation. Significant research is currently being conducted in the water industry related to concentrate treatment and volume reduction strategies to reduce the cost associated with concentrate handling. Identifying treatment technologies that can be used as an alternate to nanofiltration and reverse osmosis, such as closed-circuit desalination, can decrease the amount of brine generated and can be considered a concentrate management strategy. In general, many technologies are available for concentrate management, but only some are proven and viable. Careful consideration must be given to site-specific constraints, concentrate mitigation objections, and project economics when selecting the right approach for concentrate management.

2.5.4 *Membrane Bioreactor and Reverse Osmosis*

Using an MBR in a potable reuse scheme provides an opportunity to bridge the gap between traditional WRRF and potable reuse treatment. An MBR is a well-proven WRRF technology that combines biological degradation of nutrients and organics with particle filtration. Because reverse osmosis typically uses MF/UF as a pretreatment step, an MBR removes the need for MF/UF and can offer cost and footprint savings. However, pathogen log removal credits are currently not granted within the United States for the MBR treatment process, which has limited its implementation in potable reuse. Australia credits pathogen log removal across an MBR subject to achieving turbidity requirements. California is currently considering adopting a similar standard.

3.0 REFERENCES

Lahnsteiner, J.; Lempert, G. (2007) Water Management in Windhoek, Namibia. *Water Sci. Technol.*, **55**, 441–448.

National Research Council (2011) *Water Reuse: Expanding the Nation's Water Supply through Reuse of Municipal Wastewater*; National Research Council of the National Academies Press: Washington, D.C.

Rice, J.; Wutich, A.; Westerhoff, P. (2013) Assessment of De Facto Wastewater Reuse across the U.S.: Trends between 1980 and 2008. *Environ. Sci. Technol.*, **47**, 11099–11105.

Rice, J.; Westerhoff, P. (2014) Spatial and Temporal Variation in De Facto Wastewater Reuse in Drinking Water Systems across the U.S.A. *Environ. Sci. Technol.*, **49**, 982–989.

Rosenfeldt, E.; Boal, A. K.; Springer, J.; Stanford, B.; Rivera, S.; Kashinkunti, R. D.; Metz, D. H. (2013) Comparison of UV-Mediated Advanced Oxidation. *J. Am. Water Works Assoc.*, **105** (7), 29–33.

Schimmoller, L.; Kealy, M. J. (2014) *Fit for Purpose: The Cost of Overtreating Reclaimed Water*; WRRF 10-01; WateReuse Research Foundation: Alexandria, Virginia.

Sichel, C.; Garcia, C.; Andre, K. (2011) Feasibility Studies: UV/chlorine Advanced Oxidation Treatment for the Removal of Emerging Contaminants. *Water Res.*, **45**, 6371–6380.

Water Environment & Reuse Foundation (2016a) *Critical Control Point Assessment to Quantify Robustness and Reliability of Multiple Treatment Barriers of a DPR Scheme*, Project 13-03; Water Environment & Reuse Foundation: Alexandria, Virginia.

Water Environment & Reuse Foundation (2016b) *Guidelines for Engineered Storage Systems,* Project 12-06; Water Environment & Reuse Foundation: Alexandria, Virginia.

WateReuse Research Foundation; American Water Works Association; Water Environment Federation; National Water Research Institute (2015) *Framework for Direct Potable Reuse*; WateReuse Research Foundation: Alexandria, Virginia.

4.0 SUGGESTED READINGS

U.S. Environmental Protection Agency (2004) *Guidelines for Water Reuse*; U.S. Environmental Protection Agency: Washington, D.C.

Walker, T. (2017) *Development of an Operation and Maintenance Plan and Training and Certification for Direct Potable Reuse Systems*, Project 13-13; Water Environment & Reuse Foundation: Alexandria, Virginia.

Water Environment Research Foundation (2015) *Considering the Implementation of Direct Potable Reuse in Colorado*; Water Environment Research Foundation: Alexandria, Virginia.

8

Monitoring and Control

Ben Stanford, Paul Biscardi, Bryce Danker, Stephanie Ishii,
Wendell Khunjar, and Phil Yi

1.0 ROLE OF MONITORING, CONTROL, AND CRITICAL CONTROL POINTS IN REUSE

1.1 Overview

In both wastewater treatment and potable water treatment, a system of monitors and controls is necessary to reliably treat water to acceptable standards. Monitors are typically used to verify treatment process integrity, but can also be used to help operations teams mitigate treatment failure. Monitoring data are made actionable through controls and can be further integrated through a critical control point (CCP) strategy. However, the specific role of monitoring and control differs between water and wastewater treatment because of differences in the source water quality, variability, application of the treated product, and the sensitivity and response time required for public health protection. Likewise, for reuse applications, an alternative set of challenges and standards are faced requiring a unique toolset of monitors and controls. Therefore, it is critical that reuse systems use treatment processes and best practices that can provide protection from integrity failure. In this chapter, these requirements are further differentiated between nonpotable reuse (NPR), indirect potable reuse (IPR), and direct potable reuse (DPR).

1.2   Role of Monitors

The purpose of a monitor is to collect actionable data from the source water, treatment facility, or distribution system. Monitors and/or sensors measure a parameter of interest either continuously or intermittently, and the rate of data collection and feedback (i.e., sample turnaround time) affects the speed at which operations teams or automated systems can respond to conditions that exceed alert or alarm levels. For example, a continuous turbidity meter may be used in a reuse system to monitor for changes in source water quality or filter performance, whereas an ammonia analyzer may take and report samples every 10 to 15 minutes. In a broader sense, monitors are used to ensure process barriers are functioning as intended to mitigate risks associated with public health, asset protection, and the environment, each of which are discussed in detail in the following sections.

1.3   Critical Control Points

One strategy that can be used to assist with identifying parameters of public health concern and establishing monitors and actionable controls is the Hazard Analysis and Critical Control Point (HACCP) methodology. The HACCP is a logical, scientific process control system designed to identify, evaluate, and control hazards, which are significant for food safety, but can also have application during NPR, IPR, DPR. The purpose of an HACCP system is to put in place process controls that will detect and correct deviations in quality processes at the earliest possible opportunity. The HACCP focuses on monitoring and maintaining the barriers of treatment rather than on end-of-pipe sampling and testing. Figure 8.1. illustrates the organizational structure of HACCP and the function of a CCP. In Figure 8.1, the reverse osmosis process step is defined as a CCP for rejection of dissolved

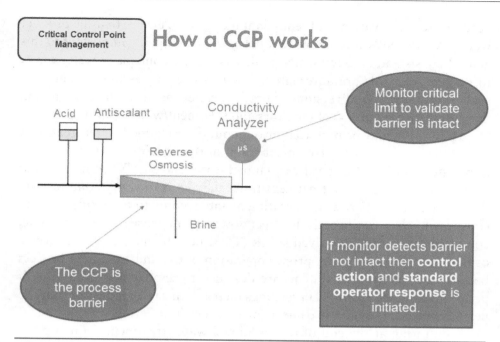

FIGURE 8.1   Description of how a CCP process barrier, monitor, and control action work together (adapted from WE&RF, 2016).

constituents. The reverse osmosis process operates as a CCP, with a critical monitor of electrical conductivity. If the electrical conductivity increases above a specific threshold, an alarm is raised and corrective action is taken by the facility operations staff or the control system can completely shut down the reverse osmosis. A detailed description of HACCP is beyond the scope of this manual, but additional information can be found in *Utilization of HACCP Approach for Evaluating Integrity of Treatment Barriers for Reuse* (Halliwell et al., 2015) and *Critical Control Point Assessment to Quantify Robustness and Reliability of Multiple Treatment Barriers of a DPR Scheme* (WE&RF, 2016).

2.0   REQUIREMENTS FOR MONITORING AND CONTROL

2.1   General Considerations

The extent of monitoring requirements in reclaimed water systems is a function of the risk associated with a water quality excursion. Stated differently, more monitoring and control is required as the likelihood of human contact or consumption increases. In high human contact reuse applications, such as residential irrigation, building humidification, and potable reuse, there is a need for frequent, varied, and redundant monitors to ensure that the product water is consistently meeting water quality targets. Equally important to this goal of consistent, targeted water quality is the use controls in water reuse. In general, controls are automated, or—in some cases—manual, actions such as shutdowns or diversions of product water that are called into action depending on the outputs of monitors and treatment/water quality setpoints.

Monitoring may be practiced throughout the reclaimed water treatment process, at the end of the treatment train, in the reclaimed water distribution system, and/or at a specific reclaimed water end use. When monitoring is practiced throughout a treatment train, the aim is to verify continuously that the treatment barrier is operating as intended or to be notified of any constituents that may negatively affect downstream processes. For example, conductivity or total dissolved solids (TDS) may be monitored in reverse osmosis permeate to ensure proper operation of the unit treatment process because reverse osmosis systems are designed to remove dissolved salts. If permeate conductivity or TDS readings increase, this is an indicator that the performance of the reverse osmosis membranes is suffering.

Monitoring at the end of the reclaimed water treatment train is practiced to assess the quality of the final product before distribution to ensure regulatory compliance and end user satisfaction. Additional monitoring in the reclaimed water distribution system and/or at a specific reclaimed water end use may be put in place to verify that there are no adverse changes in water quality during distribution and use (e.g., recirculation in industrial cooling applications).

In general, the approach to reclaimed water monitoring should be the same as that described by the National Research Council (2012) for drinking water systems, which included three parts, as follows: (1) characterization of elements that control the performance of unit processes in removing specific contaminants (e.g., empty bed contact time for granular activated carbon [GAC] treatment; low sludge density index for reverse osmosis), (2) identification of parameters that can be reliably monitored to confirm that treatment barriers are performing as expected (e.g., conductivity and TDS monitoring of reverse osmosis permeate), and (3) routine analysis of the final product to confirm that upstream measures are reliable (e.g., grab and composite samples for on-site/off-site laboratory analyses [NRC, 2012]).

2.2   Monitoring Support of Treatment Goals

As previously discussed, monitoring is practiced to verify proper operation of a treatment barrier, to protect downstream processes, and to ultimately ensure the quality of the final product. Herein, monitoring is described as being in support of one or more of the following goals: human health protection, environmental health protection, and asset management.

2.2.1 Human Health Protection

The main barriers for protection of public health are provided by the treatment processes used in a reuse facility. In reuse applications, especially as the potential for human contact with reclaimed water increases, there is an emphasis on disinfection performance and the biological stability of the product water. The focus on the biological nature of reclaimed water stems from the fact that most human contact with reclaimed water is for a short duration of time (i.e., NPR applications) and pathogens present the most potential for acute risk. In potable reuse applications, however, human health concerns become more pronounced, including parameters associated with both acute risk (e.g., pathogens) and chronic risk (e.g., trace organic compounds), and the process falls under drinking water treatment regulations with an increased emphasis on pathogen reduction and chemical reduction goals. An important part of human health protection is the implementation of an industrial pretreatment and monitoring program to minimize the risk associated with spill and/or illegal discharge events, as recycled water quality is affected by influent water quality. A pretreatment program enables publicly owned treatment works to protect their infrastructure and finished water quality from nondomestic wastewater inputs through the use of general and specific prohibitions, categorical pretreatment standards, and local limits on industrial, commercial, and institutional dischargers within the sewer collection system.

2.2.2 Environmental Health Protection

Water reuse not only presents the opportunity for human contact with reclaimed water, but also, in most cases, guarantees the opportunity for reclaimed water to come into contact with the environment. Interactions between reclaimed water and the environment may be purposeful, as is the case of irrigation, or a secondary effect, such as when runoff is produced as a result of firefighting or car washing. Similar to limits for environmental discharges of effluent to receiving waterbodies, reclaimed water must be monitored for water quality parameters that may affect environmental health. Monitoring that ensures the biologically benign nature of reclaimed water continues to be a priority for environmental health protection, as do additional parameters such as salinity with specific anion and cation limitations depending on the use, biochemical oxygen demand (BOD), dissolved oxygen, and nutrients.

2.2.3 Asset Management

A monitoring and control strategy can also play an integral role in asset management. The implementation of monitors intended to protect treatment processes and other equipment in the reclaimed water system is considered to support the goal of asset management. This type of monitoring requires the identification of water quality parameters or indices that serve as warning signs for the specific system under consideration. For example, operators of systems using GAC treatment may be interested in monitoring the influent organic load, such as through the measurement of dissolved organic carbon, UV absorbance at 254 nm, and/or carbonaceous oxygen demand (COD) via dichromate or alternative methods (Geerdink et al., 2017). Unexpected spikes in the organic content of influent water, such as that which may result from a chemical or grease input to the collection system, can significantly affect the remaining life of GAC contactors because of the exhaustion of sorption sites. Likewise, monitoring filter performance, membrane integrity, UV lamp life, and so on will help inform when various assets or parts of those assets will need to be replaced.

2.3   Examples of Regulatory Requirements and Implications

Specific examples of regulatory monitoring requirements are described in the following subsections. This discussion is not intended to serve as a comprehensive review of regulatory requirements, but rather to provide examples of how certain monitors and controls may be put in place for each monitoring goal. For a more in-depth discussion of reclaimed water regulations, refer to Chapter 4.

2.3.1   Human Health Protection

Turbidity monitoring is a common requirement for reclaimed water systems because turbidity serves as an indicator of bacterial contamination and measurements may be taken on a frequent basis. For example, the Virginia Department of Environmental Quality requires that turbidity analyses be performed by a continuous, online turbidity meter equipped with an automated data recorder if reclaimed water is used for high human contact applications (i.e., Level 1 reclaimed water). Manual samples for turbidity readings are only allowed if the online turbidity meter goes out of service. Total residual chlorine must also be continuously measured by an online monitor before Level 1 reclaimed water distribution. For both turbidity and total residual chlorine, corrective action thresholds and associated action plans are established as a control for bacterial contamination (Virginia Administrative Code, 9VAC25-740).

Similarly, the Texas Commission on Environmental Quality also requires that turbidity be measured in reclaimed water that is likely to come into contact with humans (Type I reclaimed water), along with *Enterococci* and fecal coliforms; however, monitoring is only required two times per week.

For Type II reclaimed water, which is reserved for uses where human contact is deemed unlikely, the frequency of *Enterococci* and fecal coliform monitoring is reduced to once per week and turbidity monitoring is not required because human health protection is less of a concern (Texas Administrative Code, 30TAC210-210).

As previously noted, monitoring requirements for human health protection become more nuanced in potable reuse applications. For example, in addition to monitoring parameters related to acute risks, California's Groundwater Replenishment Using Recycled Water regulations include the monitoring of indicator compounds and surrogate parameters to ensure system performance in managing chronic and acute risks. Indicator compounds are individual chemicals in the source water that represent the characteristics of a specific family of trace organic chemicals, thus allowing their monitoring to serve as a measurement of that family's removal during treatment. For chronic risk compounds, utilities can periodically collect samples and analyze for the presence of certain indicator compounds with different chemistries and removal mechanisms. Surrogate parameters are measurable properties that have a direct correlation with the concentration of one or more indicator compounds and must also be used to measure the removal of trace organic compounds in treatment (California Department of Public Health, DPH-14-003E).

2.3.2 Environmental Health Protection

Although one may not consider nutrients and oxygen demand to be directly related to human health, these parameters are critical for environmental health protection and are thus typically included in reclaimed water monitoring requirements. As evidence of their importance, the State of Washington requires that 24-hour composite samples of reclaimed water be taken on a weekly basis for the measurement of BOD, with the acceptable concentration being a function of the reuse application. At a minimum, the monthly arithmetic mean of all samples shall not exceed 30 mg/L; however, the monthly arithmetic mean limit is reduced to 20 mg/L for wetland discharge and the 7-day arithmetic mean limit is further reduced to 5 mg/L for direct aquifer discharge because of the sensitivity of these receiving waters. In reuse applications in the state of Washington involving the discharge of reclaimed water to wetlands, 24-hour composite samples must also be collected at least weekly to ensure compliance with respective total nitrogen, total Kjeldahl nitrogen (TKN), ammonia, and phosphorus limits (WDOE, 1997).

Similarly, the Virginia State Water Control Board adopted two policies to regulate effluent discharges to receiving waters that serve as public drinking water supplies: the Occoquan Policy (Virginia Administrative Code, 9VAC25-410) and the Dulles Area Watershed Policy (Virginia

Administrative Code, 9VAC25-401). These policies recognize this effluent management strategy as a form of planned potable reuse, often referred to as *surface water augmentation,* and the need to produce high-quality effluent for the protection of human and environmental health. To this end, water resource recovery facilities discharging to the Occoquan and Dulles Area watersheds are required to meet advanced treatment and stringent effluent requirements, such as monthly average effluent limits of 10.0 mg/L COD, 1.0 mg/L TKN, 0.1 mg/L total phosphorus, and 0.5 NTU turbidity.

2.3.3 Asset Management

Monitoring for asset management tends to be implemented as part of an overall asset management strategy. Asset management can help improve the function, production, and overall cost-effectiveness of a facility. Monitors for asset management may include tracking the age of assets, the amount of power to operate equipment, and the allowable breakthrough of given parameters (e.g., irreversible flux decline in membranes, total organic carbon [TOC] or COD breakthrough in GAC), and should be decided on with guidance from asset management specialists.

3.0 TRENDS IN REAL-TIME MONITORING AND CONTROL

3.1 Overview and Trends in Real-Time Monitoring and Control

Monitors can provide a means to verify the performance of barriers used in a treatment process scheme (e.g., CCPs for managing human health risks) or they can be used to verify that finished water meets water quality objectives and compliance goals. To that end, monitors that are used to validate that a process is functioning properly (e.g., a filter effluent turbidity monitor) may not necessarily tell the operator *directly* that pathogens are being removed, rather, it provides a surrogate parameter (e.g., turbidity) that informs the operator that, within a certain set of limits, the desired water quality goals will be met. Considering that the U.S. Environmental Protection Agency's definition of safe drinking water would require less than 2.2×10^{-7} viruses per liter, a drinking water facility would need to process more than 1 200 000 gal of water through an analyzer to find that one pathogen. This obviously becomes infeasible and, therefore, the focus needs to be on surrogate parameters such as turbidity, disinfectant residual, chlorine concentration multiplied by time (C*T), and so on.

Chemical constituents, on the other hand, tend to be homogenously distributed throughout a water sample and, therefore, detection becomes more straightforward, but may be limited by sensitivity of analytical methods. As such, there is a significant push in the industry to develop sensors capable of a high degree of sensitivity while maintaining a high throughput. Additional efforts are focusing on the development and validation of bioassays to determine the safety of reclaimed water. While many bioassays currently exist, the state of the science does not allow the research community to equate a specific bioassay endpoint with a defined risk to human health. Thus, significant research is needed to convert bioassays into actionable monitoring strategies for reclaimed water and drinking water utilities.

3.2   Selection of Monitors for Process Function and Control

Although several projects are ongoing in their evaluation of monitors for water reuse, guidance is available from *Critical Control Point Assessment to Quantify Robustness and Reliability of Multiple Treatment Barriers of a DPR Scheme* (WE&RF, 2016) and *Development of an Operation and Maintenance Plan and Training and Certification for Direct Potable Reuse* (WE&RF, 2017) for selecting monitors in potable reuse applications. When there is a risk that a critical process may fail, proper monitoring must be in place to inform operators about the functioning of that process and when changes may need to be made to that asset. The example shown in Table 8.1, adapted and truncated from *Critical Control Point Assessment to Quantify Robustness and Reliability of Multiple Treatment Barriers of a DPR Scheme* (WE&RF, 2016), provides one option for organizing information about CCP monitors for potable reuse. The table is divided up into CCPs, the associated health risk that CCP is protecting against ("Risk Parameter"), the analyzer or monitor associated with demonstrating acceptable performance of the CCP, and reasons for a potential failure of the barrier ("Trigger Point/ Failure Mode").

3.3   Assessing Monitor Reliability

Recent research published by the Water Environment & Reuse Foundation (WE&RF, 2016) provided an analysis of monitors and their reliability for DPR, and the methodology therein can be used to identify where redundant monitoring should be implemented to maximize public health protection.

When there is a risk that a critical process may fail, proper monitoring must be in place to inform operators about the functioning of that process and when changes may need to be made to that asset. However, there is also a risk that a monitor and, subsequently, the operator, will fail to notice the failure or improper operation of a specific control point. In that sense, the risk of "failure to notice failure" needs to be quantified to understand how likely such an outcome would be and to assess the extent of effect such a failure would cause.

TABLE 8.1   Example organization of process-monitor pairings for an ozone-biofiltration-based potable reuse system (adapted from WE&RF, 2016).

Critical control point	Risk parameter	Analyzer/ monitor	Trigger point/failure mode CCP
Ozone	Pathogens Organic chemicals Bromate	UV transmittance analyzer Ozone dose analyzer	Insufficient dose Overdose
Ozone-biologically active carbon (BAC)	Biological stability disinfection byproducts (DBPs) precursors	Ozone BAC dose analyzer Magnetic flow analyzer	Insufficient dose Insufficient contact time with BAC
Coagulant-BAC	Pathogens	TOC analyzer Flow meter analyzer	Insufficient coagulant dose Filter breakthrough
GAC	Organic contaminants	TOC analyzer UV analyzer	Carbon too old Filter bypass
UV	Pathogens	UV transmissivity analyzer	Insufficient UV dose Poor transmissivity
Chlorine	Pathogens DBP formation Organic chemicals	Chlorine analyzer	Insufficient dose (dosing pump failure)

When considering failure, the risk priority number (RPN) can be used to evaluate the likelihood and effect of a failure event for a monitor (WE&RF, 2016). In short, the occurrence (O), severity (S), and detectability (D) of monitor failure is ranked according to specific guidelines on a scale of one to 10, with one indicating infrequent occurrence, minimal effect, and a high likelihood of detecting the issue, and 10 indicating a high degree of occurrence or effect and low likelihood of detecting the event. Those scores are then multiplied together (as shown in Table 8.2) to calculate the RPN, whereby a higher score would indicate a greater risk and effect from monitor failure. Design teams would then use the RPN to determine where additional redundant monitors may be required and/or where more frequent maintenance, calibration, and validation of monitors should be implemented. Although no specific guidance is provided on actions related to a specific score, values over the 100- to 150-point range should be evaluated

TABLE 8.2 Example calculation of risk priority number for chlorine and ammonia analyzers.

Component name	Cause(s) of failure	Effect(s) of failure	Failure mode(s)	O	S	D	Risk priority number (O)*(S)*(D)	Recommended corrective action
Chlorine analyzer	Over-dose	Disinfection byproduct formation	Chlorine analyzer reads false low result, leading to overdose	4	6	4	96	Cross-check reading, calibrate instrument and replace probe if necessary
Ammonia analyzer	Incorrect dose	Disinfection byproduct formation	Ammonia analyzer reads false high result, leading to overdose	2	5	4	40	Cross check reading, calibrate instrument and replace probe if necessary

for potential needs of adding redundant monitoring. Monitor redundancy may involve duplicate monitors for the same parameter or the use of monitors that measure multiple surrogate parameters, each of which indicate the proper functioning of a given unit process.

4.0   IMPLICATIONS AND OPPORTUNITIES FOR INNOVATION

As noted previously, there are several key implications related to monitoring and control for water reuse. These are as follows:

- Public health,
- Product water quality,
- Capital and operational costs, and
- Operational complexity and personnel.

Ensuring product water quality is critical, but it is achieved at a cost and degree of operational complexity. Consequently, capital and operation and maintenance costs cannot be ignored. Process monitoring and control can play a critical role in ensuring that product water quality is maintained, risks are mitigated, and costs are controlled. Advanced process monitoring can help inform baseline performance and facilitate development of control strategies that reduce operating costs. In addition to benchmarking treatment performance, process monitoring and control in distribution systems is necessary to ensure that high-quality product water is delivered to the consumer.

As the complexity of monitoring and control increases, data generation will also increase. There exists an opportunity to develop decision frameworks that assist staff with understanding what parameters/controls are truly critical to maintaining compliance. Data management strategies need to be aligned with new training for personnel. Current wastewater and drinking water staff curricula need to be amended to ensure that operational and maintenance staff are appropriately cross-trained to be able to understand and identify potential risks, mitigation and remediation strategies, and metrics for long-term performance monitoring.

Online instrumentation is often able to provide real-time data monitoring, which is important; however, managing, archiving, and then displaying the collected data in a manner that is useful for the operator is of equal importance. Effective data management and displays through development of dashboards can help personnel quickly understand performance and also potential issues that may occur. Although each system is unique, there exists an opportunity for the industry to develop guidelines for helping utilities establish effective data management strategies that fulfill specific needs while allowing comparison of performance among peer utilities. As online instrumentation continues to grow along with technology, cloud-based systems, and online networking, cyber security of water reuse systems and protecting the data management networks will be critical. It is necessary for organizations to assess their level of cyber risk and determine the appropriate security measures for their specific systems. Protecting networks is no longer optional, but absolutely necessary.

There also exist opportunities for innovative monitoring and control strategies. Advances in online instrumentation for monitoring and control for both water and wastewater can be effectively used to inform decisions regarding water reuse. Monitoring the water quality of the environmental buffer can be important because different environmental buffers will have varied water quality characteristics. Online instrumentation can help with developing dynamic control strategies to understand when bypass may be required, blending, equalization or diversion of off-spec water, as well as total residual chlorine within a water reuse distribution system. One example of the use of online instrumentation is at the Orange County Water District's Groundwater Replenishment System (California), where online analyzers are used to continuously monitor and track the performance of the reverse osmosis system.

Within this context, there is room to improve the use of fluorescence and/or UV-based sensors for monitoring dissolved organic matter. There has been some improvement of these types of analytical tools (e.g., new chemical oxygen demand analyzers); however, there are still limitations and a need for further development and tuning of these analyzers and sensors. Monitoring of dissolved organics and nutrients is critical for understanding the biological stability of reclaimed water. Improved monitoring capabilities of parameters such as assimilable organic carbon or biodegradable dissolved organic carbon would help in reducing the associated error in current measurement techniques and analytical costs.

As data collection and generation increases, there is a need for innovation in the development and use of fault-detection algorithms to help with processing large amounts of data. Adoption of these algorithms/methods can help with predictive modeling for the early detection of failure (e.g., principal component analysis) and help engineers and operators preserve quality of operations.

There is also a need to develop and use tools that can provide integrated modeling of entire water systems (i.e., wastewater effluent to environmental buffer or raw water intake). These tools can be used to help refine design as well as optimize operation. Once developed, these models can serve as the platform for training of operations staff.

5.0 REFERENCES

Geerdink, R. B.; van den Hurk, R. S.; Epema, O. J. (2017) Chemical Oxygen Demand: Historical Perspectives and Future Challenges. *Anal. Chim. Acta,* **961,** 1–11.

Halliwell, D.; Burris, D.; Deere, D.; Leslie, G.; Rose, J.; Blackbeard, J. (2015) *Utilization of HACCP Approach for Evaluating Integrity of Treatment Barriers for Reuse*; WateReuse Research Foundation: Alexandria, Virginia.

National Research Council (2012) *Water Reuse: Potential for Expanding the Nation's Water Supply Through Reuse of Municipal Wastewater*; The National Academies Press: Washington, D.C.

Water Environment & Reuse Foundation (2017) *Development of an Operation and Maintenance Plan and Training and Certification for Direct Potable Reuse (DPR) Systems*, Project 13-13; Water Environment & Reuse Foundation: Alexandria, Virginia.

Water Environment & Reuse Foundation (2016) *Critical Control Point Assessment to Quantify Robustness and Reliability of Multiple Treatment*

Barriers of a DPR Scheme, Project 13-03; Water Environment & Reuse Foundation: Alexandria, Virginia.

Washington State Department of Ecology (1997) Water Reclamation and Reuse Standards; Publication No. 97-23; Washington State Department of Ecology: Lacey, Washington.

9

Ongoing Maintenance and Monitoring Progress

Christopher Stacklin, P.E., and Bruce L. Cooley, P.E.

1.0 OVERVIEW OF ONGOING MAINTENANCE AND UPKEEP

Water resource recovery facilities (WRRFs) are no longer the traditional facilities of the past. In consideration of the newer role of fitful repurposing of water for beneficial reuse, as discussed by Schimmoller and Kealy (2014) in Chapter 7, monitoring metrics for maintenance and upkeep of capital assets have been elevated to higher standards than required previously. As utility management looks for ways to optimize efficiency, maximize recovery of water, and contribute to the overall health of the watershed, proactive approaches must be executed. This includes a commitment to long-term sustainability of WRRFs achieved by proactive and upfront investment in system resiliency and innovation.

Maintenance management is vital to stewarding WRRF capital improvement program budgets. Based on a limited study in California, the Pacific Institute estimated the leveled facility cost, or the cost per unit that results in a break-even investment, of indirect potable reuse (DPR) facilities with an annual production less than 33 794 m^3/d (10 000 ac-ft/yr [8.92 mgd]) ranges from $1.22 to $1.78/m^3 ($1,500 to $2,200/ac-ft [$4.60 to $6.75/1000 gal]), with a median cost of $1.54/m^3 ($1,900/ac-ft [$5.83/1000 gal]). The cost of larger projects ranges from $0.89 to $1.30/m^3 ($1,100 to $1,600/ac-ft [$3.38 to $4.91/1000 gal]), with a median cost of $1.05/m^3 ($1,300/ac-ft [$3.99/1000 gal]) (Cooley and Phurisamban, 2016). In *Framework for Direct Potable Reuse*, the treatment cost for an advanced water treatment facility with reverse osmosis costs $0.56 to $0.73/m^3 ($685 to $900/ac-ft [$2.10 to $2.76/1000 gal]) (Tchobanoglous et al., 2015). For reference, the $0.56/m^3 ($685/ac-ft [$2.10/1000 gal]) is based on the Orange County Water District Advanced Water Treatment Facility (California), which has an installed maximum capacity of 492 104 m^3/d (130 mgd) of produced potable water and has been in service since 2008. The facility is configured for full advanced treatment (FAT), which includes microfiltration and reverse osmosis, followed by an advanced oxidation process.

The Texas Water Development Board commissioned a study in 2015 that compared the capital and maintenance costs of six difference water reuse treatment schemes, ranging in capacity from 3785 to 94 635 m^3/d (1 to 25 mgd). By expanding the data over a 20- and 30-year life cycle of a water reuse facility, operations and maintenance represents between 39 and 82% of the water reuse facility life cycle cost (Texas Water Development Board, 2015).

As a comparison, in 2004, the U.S. Environmental Protection Agency (U.S. EPA) estimated the annual O&M cost for water reuse at 21% based on an operation and maintenance (O&M) cost at $0.08 to $0.39/m^3 ($100 to $478/ac-ft, or $0.31 to $1.47/1000 gal) without federal and state reimbursements (U.S. EPA and U.S. AID, 2004).

Energy is often attributed as one of the largest O&M expenses, accounting for 30 to 55% of the O&M costs (Cooley and Phurisamban, 2016). This is often more than labor costs, a relation noted in the Congressional Research Service report, *Energy-Water Nexus: The Water Sector's Energy Use* (Copeland and Carter, 2017).

Because O&M activities are strongly related to energy, the roadmap for maintenance can be based on making the best of energy demand for the water reuse facility, with opportunity for energy integration between the water reuse and wastewater treatment sides of the WRRF (U.S. EPA, 2013). Possible integration would be expanding energy recovery of the wastewater treatment side via co-digestion to increase electric power production for the water reuse side (Nghiem et al., 2017). Opportunities to raise digester gas on the wastewater treatment side to offset energy consumed by the advanced treatment unit processes are a feasible consideration that would also improve resiliency (Milbrandt et al., 2016) and reduce dependency on outside electric power. Optimization considerations can focus on energy recovery of high-energy streams. For example, part of the energy stored in reverse osmosis concentrate can be recovered using microturbines (Raucher and Tchobanoglous, 2014). Facilities based on reverse osmosis unit processes are testing low-pressure membranes to reduce energy consumed by pumping across the reverse osmosis membranes (Knoell et al., 2016). The Water Environment & Reuse Foundation, in partnership with WateReuse California, launched the Direct Potable Reuse Initiative in June 2012 to advance DPR as a water supply option in California. Several research projects have been initiated that explore low-energy treatment for potable and nonpotable reuse projects (Mosher et al., 2016).

Because O&M costs are a significant portion of life cycle costs, use of reliability-centric asset management programs is justified to aid in best management of limited resources (Cooley, 2016; Marlow and Burn, 2008). To return the best values to rate-paying customers, on-stream or on-demand water production becomes important to optimize. Staged and shadowed maintenance strategies can be used as part of the on-stream water production optimization process, which has been well developed in the power industry (Stacklin, 1989, 1990).

In the case of FAT, multiple trains of reverse osmosis units are operated in parallel. Variable water demand results in changes in the spare capacity of the units. By predicting the water demand and, thereby, which units can be designated as a spare, staged maintenance can be pre-planned to take down specific reverse osmosis trains to perform maintenance without affecting on-stream water demand. In the case of shadowed maintenance, taking a reverse osmosis train down for maintenance allows for units or subunits within a train to be maintained. For example, taking a reverse osmosis train offline for maintenance allows for forwarding pumps upstream of the reverse osmosis units to be serviced because water production is at reduced capacity.

Predictive maintenance can also optimize on-stream water production as part of a maintenance management approach. Predictive maintenance uses monitoring technology or other available data to estimate when equipment operations may adversely affect system operations. To establish reasonable and realistic key performance indicators for system performance, system data must be collected, analyzed, and routinely evaluated. In the water reuse industry as a whole, these data are not often pooled or shared.

Pooling and sharing reliability data can have a positive benefit on the efficient execution of maintenance activities. A model for pooling data for energy-related systems is the North American Electric Reliability Corporation (NERC). The NERC is a not-for-profit, international regulatory authority whose mission is to ensure the reliability and security of the bulk power system in North America. The NERC develops and enforces reliability standards and annually assesses seasonal and long-term reliability. The data compiled by NERC can be used for both reliability-centered maintenance, condition assessment, asset management, and design for future water reuse facilities.

2.0 INFORMING PROGRESS TO PUBLIC AND REVISITING THE NEEDS STATEMENT

The triple-bottom-line concept is associated with social, environmental, and economic cost. This approach aids the users in considering both qualitative and quantitative criteria. The triple-bottom-line approach can be used to identify a pricing level structure that supports water reuse systems (WEF, 2016). With a triple-bottom-line approach to maintenance management, water reuse utilities are obligated to examine ongoing social and environmental effects and not solely O&M costs. Two primary social and environmental effects to consider are maintaining the public trust and environmental justice.

From the perspective of public trust with maintenance and upkeep, having a robust monitoring and reporting program (MRP) is a continuous validation of compliance of water reuse technologies, and is well understood from the science perspective. However, despite having MRP data that document the ability of water reuse technologies to consistently treat municipal wastewater effluent to meet or exceed drinking water quality standards, many utilities face public skepticism.

Relatively new approaches to building public trust and fostering adoption of water reuse is based on the concepts of societal legitimacy, which is the generalized perception or assumption that a technology is desirable or appropriate within the social context (Binz et al., 2016; Harris-Lovett et al., 2015). Integration of the water reuse project within the community becomes important on a continuous basis, gauged by surveys and studies (Ishii et al., 2015). This not only fosters trust in the water reuse approach, but also builds trust in the governing agency. In this way, societal legitimacy becomes an ongoing and essential process component of maintenance.

Another important effect to consider in the triple-bottom-line approach is environmental justice, which is defined as the fair treatment and meaningful involvement of all people regardless of race, color, national origin, or income, with respect to the development, implementation, and enforcement of environmental laws, regulations, and policies (Beveridge et al., 2017; Schlosberg, 2009; U.S. EPA, 2017). Environmental justice has a nexus with water reuse and is an ongoing process that influences the price and quality of the produced water. Although the U.S. EPA sets the national drinking water quality standard, supplemented by state and regional standards and objectives, produced water from reuse facilities is uniform, that is to say, the produced water conforms to the most stringent standard and is distributed to all levels of society in the same level of quality.

Pricing or rate structures can be framed with environmental justice and societal legitimacy. The cost of produced water affects the community served by the supply; however, the cost effect can be offset by resiliency and reliability of a quality water supply portfolio.

Periodic evaluation of societal legitimacy and environmental justice can provide perspective on the reuse facility needs statement and expand public awareness of non-monetarized value within the community, thus building public trust and confidence in the produced water from the reuse facility.

3.0 REFERENCES

Beveridge, R.; Moss, T.; Naumann, M. (2017) Sociospatial Understanding of Water Politics: Tracing the Multidimensionality of Water Reuse. *Water Alternatives*, **10** (1), 22–40.

Binz, C.; Harris-Lovett, S.; Kiparsky, M.; Sedlak, D.; Truffer, B. (2016) The Thorny Road to Technology Legitimation—Institutional Work for Potable Water Reuse in California. *Technological Forecasting and Social Change*, **103**, 249–263.

Cooley, B. L. (2016) Financially Stressed? Maybe There Is a Better Way; Paper presented at the Indiana Water Environment Association Specialty Conference; Indianapolis, Indiana, Oct 13.

Cooley, H.; Phurisamban, R. (2016) *The Cost of Alternative Water Supply and Efficiency Options in California*; Pacific Institute: Oakland, California.

Copeland, C; Carter, N. T. (2017) *Energy-Water Nexus: The Water Sector's Energy Use*; Congressional Research Service 7-5700 R43200; Congressional Research Service: Washington, D.C.

Harris-Lovett, S. R.; Binz, C.; Sedlak, D. L.; Kiparsky, M.; Truffer, B. (2015) Beyond User Acceptance: A Legitimacy Framework for Potable Water Reuse in California. *Environ. Sci. Technol.*, **49** (13), 7552–7561.

Ishii, S.; Boyer, T.; Cornwell, D.; Via, S. (2015) Public Perceptions of Direct Potable Reuse in Four U.S. Cities. *J. Am. Water Works Assoc.*, **107** (11), E559-E570.

Knoell, T.; Gonzalez, R.; Carter, S.; Guibert, S.; Stefanic, M. (2016) Evaluating the Latest Low-Pressure Membrane Technologies for the Groundwater Replenishment System. *Proceedings of the 89th Annual Water Environment Federation Technical Exposition and Conference* [CD-ROM]; New Orleans, Louisiana, Sep 24–28; Water Environment Federation: Alexandria, Virginia.

Marlow, D. R.; Burn, S. (2008) Effective Use of Condition Assessment within Asset Management. *J. Am. Water Works Assoc.*, 100, 1, 54–63.

Milbrandt, A.; Bush, B.; Melaina, M. (2016) *Biogas and Hydrogen Systems Market Assessment*; No. NREL/TP-6A20-63596; National Renewable Energy Laboratory: Golden, Colorado.

Mosher, J. J.; Vartanian, G. M.; Tchobanoglous, G. (2016) *Potable Reuse Research Compilation: Synthesis of Findings*; WRRF-15-01; Water Environment & Reuse Foundation: Alexandria, Virginia.

Nghiem, L. D.; Koch, K.; Bolzonella, D.; Drewes, J. E. (2017) Full Scale Co-Digestion of Wastewater Sludge and Food Waste: Bottlenecks and Possibilities. *Renew. Sustain. Energy Rev.,* 72, 354–362.

Raucher, R. S.; Tchobanoglous. G. (2014) *The Opportunities and Economics of Direct Potable Reuse*; WateReuse Research Foundation: Alexandria, Virginia.

Schimmoller, L.; Kealy, M. J. (2014) *Fit for Purpose: The Cost of Overtreating Reclaimed Water*; WRRF 10-01; WateReuse Research Foundation: Alexandria, Virginia.

Schlosberg, D. (2009) Defining Environmental Justice: Theories, Movements, and Nature; Oxford University Press: Oxford, U.K.

Stacklin, C. (1989) Guaranteeing Availability in First-of-a-Kind Plants: The Midland Experience. Paper presented at the American Society of Mechanical Engineers Joint Power Generation Conference, 89-JPGC/Pwr-6; Dallas, Texas, Oct 22–26.

Stacklin, C. (1990) Pairing On-Line Diagnostics with Real-Time Expert Systems. *Power,* **134** (6), 55–58.

Tchobanoglous, G.; Cotruvo, J.; Crook, J.; McDonald, E.; Olivieri, A.; Salveson, A.; Trussell, S. R. (2015) *Framework for Direct Potable Reuse*; WateReuse Research Foundation: Alexandria, Virginia.

Texas Water Development Board (2015) Direct Potable Reuse Resource Document Final Report; TWDB Contract No. 1248321508, Volume 1 of 2, April; pp 5-9–5-19.

U.S. Environmental Protection Agency (2013) *Energy Efficiency in Water and Wastewater Facilities, A Guide to Developing and Implementing Greenhouse Gas Reduction Programs*; U.S. Environmental Protection Agency: Washington, D.C., pp 2–5.

U.S. Environmental Protection Agency (2017) Environmental Justice. https://www.epa.gov/environmentaljustice (accessed May 2017).

U.S. Environmental Protection Agency; U.S. Agency for International Development (2004) Chapter 6, Funding Water Reuse Systems. In *Guidelines for Water Reuse*; EPA-625/R-04-108; U.S. Environmental Protection Agency: Washington, D.C.

Water Environment Federation (2016) *The Water Reuse Roadmap Primer*; Water Environment Federation: Alexandria, Virginia.

Index

CPSIA information can be obtained
at www.ICGtesting.com
Printed in the USA
LVHW021738191220
674585LV00001B/10